ENGINES OF
DISCOVERY

A Century of Particle Accelerators

ENGINES OF DISCOVERY

A Century of Particle Accelerators

Andrew Sessler

Lawrence Berkeley National Laboratory, USA

Edmund Wilson

CERN, Geneva, Switzerland

 **World Scientific**

NEW JERSEY · LONDON · SINGAPORE · BEIJING · SHANGHAI · HONG KONG · TAIPEI · CHENNAI

Published by

World Scientific Publishing Co. Pte. Ltd.

5 Toh Tuck Link, Singapore 596224

USA office: 27 Warren Street, Suite 401-402, Hackensack, NJ 07601

UK office: 57 Shelton Street, Covent Garden, London WC2H 9HE

Library of Congress Cataloging-in-Publication Data
Sessler, A. M. (Andrew Marienhoff)
     Engines of discovery / a century of particle accelerators / Andrew Sessler, Edmund Wilson.
         p. cm.
     Includes bibliographical references and index.
     ISBN-13 978-981-270-070-4 -- ISBN-10 981-270-070-6
     ISBN-13 978-981-270-071-1 (pbk.) -- ISBN-10 981-270-071-4 (pbk.)
   1. Particle accelerators. 2. Particle accelerators--Design and construction. I. Wilson, E. (Edmund)

QC787.P3 S47 2007
539.7'3--dc22

                                                                                          2007060671

British Library Cataloguing-in-Publication Data
A catalogue record for this book is available from the British Library.

Cover picture: 'Propellers' by Fernand Leger. © 2006 ProLitteris, Zurich.

Printed by Mainland Press Pte Ltd

Preface

Much of the raw material for this book was collected from the work of others, sometimes from carefully constructed review documents, sometimes from books, and often simply from memories of conversations with our colleagues over the last 50 years. This is not a very technical book but we hope that we have recorded as much as possible of the history of particle accelerators — the Engines of Discovery — as well as the lives of those that built them, before it is either forgotten or lost to living memory.

This work is a personal perspective and, apart from the kind of errors that are inevitable in a first printing, we must surely have omitted to mention incidents and personalities that the reader would have liked to have seen included. We hope that further editions will rectify this.

The book contains many sidebars — biographical notes and descriptions of laboratories as well as technical concepts. Writing the biographical sidebars in this book would not have been possible without the generous help of our colleagues. Nevertheless, lack of space and the difficulty in tracking down reliable sources of information has meant that we have had to make a rather arbitrary and subjective choice. We have had to leave out many accomplished individuals who have made important contributions to the field — and regrettably risk disappointing some of our personal friends. The names of those who have been left out may be found in the appendices which include a list of those who have been awarded prizes by their professional bodies — others appear in our list of principal publications. The appendices also include a glossary of commonly used technical terms and abbreviations together with a bibliography listing general texts, more technical accelerator books, seminal publications, and some web addresses.

We wish to thank David Whittum and Dieter Möhl for reading through an early draft and making many useful suggestions and we are very grateful to Jose Alonso, André Anders, André Barlow, Joe Chew, Tim Houck and Stefano De Santos for their hard work in correcting the proofs.

We are thankful to the Lawrence Berkeley National Laboratory and to CERN for supporting our activities. One of us (AMS) was supported by the US Department of Energy, Office of Basic Energy Sciences, under Contract No. DE-AC02-05CH11231.

Accelerator science and engineering — the ability to handle particle and photon beams — has developed far beyond what anyone could have imagined in 1900. This book is dedicated to the physicists and engineers who, over a period of 100 years made possible electrostatic machines, the cyclotron, the betatron, the linac, the synchrotron, colliders and the other machines. These machines have discovered new forms of matter and changed our lives for the better.

Contents

Preface   v

Introduction   xi

Chapter I. Electrostatic Accelerators   1

I.1	Scientific Motivation	1
I.2	Voltage Multiplying Columns	4
I.3	Silk Belts	5
I.4	Wisconsin Advances	6
I.5	Tandems	8
I.6	Commercial Production of Electrostatic Machines	9
I.7	Applications of Electrostatic Machines	9

Chapter II. Cyclotrons   11

II.1	The Anatomy of a Discovery	11
II.2	Lawrence and the Early Cyclotrons	14
II.3	Transverse Focusing	18
II.4	Relativistic Limitation	19
II.5	Calutrons	19
II.6	Cyclotrons for Peace Again	21
II.7	FFAG	22
II.8	Spiral Sector Cyclotrons	23
II.9	Modern Cyclotrons	26
II.10	Applications	26

Chapter III. Linear Accelerators   28

III.1	Science Motivation — An Idea in Search of a Technology	28
III.2	The Early Linear Accelerators at Berkeley	30
III.3	Proton Linacs	30
III.4	Electron Linacs	34
III.5	Heavy Ion Linacs — a Rich Field of Research	42
III.6	Induction Linacs	44
III.7	Applications of Induction Linacs	46

Chapter IV. Betatrons   49

IV.1   Early History   50
IV.2   The Kerst Betatron   51
IV.3   The Wideroe Betatron — Second Attempt   51
IV.4   The Years After World War II   53

Chapter V. Synchrotrons   55

V.1   Science Motivation   55
V.2   The Early History of the Synchrotron   55
V.3   First Synchrotron   56
V.4   Electron Synchrotrons   56
V.5   Early Proton Synchrotrons   56
V.6   Nimrod and Phasotron   60
V.7   Strong Focusing   60
V.8   Brookhaven's AGS and CERN's PS   64
V.9   Fermilab and SPS   67
V.10   Superconducting Magnets   75

Chapter VI. Colliders   78

VI.1   Science Motivation   78
VI.2   Principles   78
VI.3   Electron–Electron Colliders   81
VI.4   Electron–Positron Colliders   81
VI.5   Superconducting Cavities   90
VI.6   Proton–Proton Colliders   91
VI.7   Proton–Antiproton Colliders   94
VI.8   Asymmetric Collider Rings   97
VI.9   Large Hadron Collider (LHC)   103
VI.10   Heavy-Ion Colliders   107

Chapter VII. Detectors   110

VII.1   Early Primitive Detectors   110
VII.2   Scintillators, Photomultipliers and Cerenkov Counters   110
VII.3   Collisions in Three Dimensions   111
VII.4   A Modern Detector   115
VII.5   Digital X-ray Imaging   117
VII.6   Detection Techniques for Synchrotron Radiation Sources   119

Chapter VIII. Synchrotron Radiation Sources   121

VIII.1   Scientific Motivation   121
VIII.2   Principles and Early History   122
VIII.3   Synchrotron Radiation   123
VIII.4   First Generation Synchrotron Sources   123
VIII.5   Second Generation Synchrotron Sources   123
VIII.6   Third Generation Synchrotron Sources   124
VIII.7   Angstrom Wavelength Free Electron Laser Facilities   128
VIII.8   Future Fourth Generation Synchrotron Sources   132

Chapter IX. Cancer Therapy Accelerators — 134

IX.1   Cyclotrons   134
IX.2   Linacs   135
IX.3   Synchrotrons   136
IX.4   Other Therapies   138
IX.5   Future Facilities   138

Chapter X. Past, Present and Future — 140

X.1   Future Needs   140
X.2   Linear Colliders and Their Origins   141
X.3   The International Linear Collider (ILC)   144
X.4   The Compact Linear Collider (CLIC)   146
X.5   Spallation Neutron Sources   148
X.6   Rare Isotope Accelerators   151
X.7   Neutrino Super Beams, Neutrino Factories and Muon Colliders   153
X.8   Accelerators for Heavy Ion Fusion and for Creating High Energy Density Plasmas   157
X.9   Proton Drivers for Power Reactors   160
X.10   Lasers and Plasmas   161

Chapter XI. A Final Word — 166

XI.1   Understanding the Universe   166
XI.2   Applications   166
XI.3   Bringing Nations Together   166
XI.4   A Word Especially for the Young   168

Appendices

A. Bibliography and References   169
B. The Accelerator Community   174
C. Glossary   175
D. List of Illustrations with Acknowledgments   184

Index — 189

Introduction

The appetite of particle physicists for particles of higher and higher energy seems never to be satisfied. Over the last century, many generations of accelerators have been built to provide particles of ever higher energy, for experiments designed to answer the latest questions concerning the ultimate structure of matter. Each leap in energy gives a rich and fascinating insight into the sub-nuclear world, showing detail never seen before, producing new particles that confirm theories which unify the forces of nature, and taking us back in time ever closer to the moment of the creation of the universe. Inevitably a deeper understanding hints at more discoveries to be made at even higher energy.

This continual search for new particles and the structures within them is no mere stamp collecting. Each discovery of a new particle stimulates the human imagination to speculate how it may fit a pattern — a pattern which can only accurately be described by an often abstruse mathematical theory connecting the forces of nature. Probing deeply into their structure reveals more of the pattern, suggesting simplification of the theory — often in an unexpected way. Each time this happens, theoretical physicists — their imagination stimulated by the challenge of explaining a new pattern — experience a flash of insight into scientific laws that describe the forces of nature: an insight that can only be tested at even higher energies. This interplay between experiment and theory is the path the human mind must follow towards a more complete understanding of nature and is the very stuff of science. "Is there more detail to be seen? Are there more particles to be discovered? If only we had higher energy!" is the refrain that follows every new verse.

More than 90 years ago, the great physicist Ernest Rutherford peered for the first time into the structure of the atom. No one had the slightest idea where the electrons, protons and neutrons were in the atom or how they were stuck together. There was no chance that optical microscopes — however powerful — might be developed to see atoms. Atoms are about 100 times smaller than the wavelength of visible light, and a light wave simply swirls past them, its wave-crests undisturbed by the tiny obstacles. Instead of a light beam Rutherford used alpha particles, nuclei of helium escaping from radioactive decay. He spent long hours in a wooden hutch, straining his eyes to see faint flashes of light on a fluorescent screen as these "alphas" were scattered by the atoms in a metal target. His observations showed an atom to be a tiny hard nucleus surrounded by electrons at a much greater distance. (See sidebar for E. Rutherford.) Reporting his exciting results to the Royal Society in London, he emphasized the limitation of using particles from radioactive decay and urged fellow scientists and electrical engineers to invent a reliable machine to accelerate more particles to higher energies.

In this book we tell the story of the particle accelerators that followed this plea — some of them the largest scientific tools known to man. It is an endeavor spanning the whole of the twentieth century. It challenged the dedication, ingenuity, and imagination of both physicists and engineers in very much the same way as had the mammoth projects and the technological triumphs — the railways, bridges, steamships and automobiles — of the nineteenth century.

We hope that even those with but a passing interest in the modern world of science will be fascinated by this story. It covers the transition of experimental science from the inventor's kitchen table to the international laboratories of today, with their thousands of employees and budgets of hundreds of millions of dollars. How did this evolve and what does it take to become one of the innovators responsible for so much progress? Among these pages we pause from time to time to give some insight into the personalities of the "accelerator people." We believe that the reader will be interested in the variety of nationality, family, upbringing, and schooling that has produced these great scientists. We

Ernest Rutherford (1871–1930)

New Zealand physicist
Nobel Prize in Chemistry, 1908
The father of nuclear science
President of the Royal Society, 1925–1930
Baron Rutherford of Nelson, New Zealand, and Cambridge

Ernest Rutherford was born, 1871, in Nelson, New Zealand, the fourth child in a family of twelve. His father, James Rutherford, a Scottish wheelwright, emigrated to New Zealand with Ernest's grandfather and the whole family in 1842. His mother was an English schoolteacher, who also went to live there in 1855.

Ernest received his early education in Government schools and at the age of 16 attended Nelson Collegiate School. In 1889 he was awarded a University scholarship to the University of New Zealand, Wellington, where he entered Canterbury College. He graduated M.A. in 1893 with a double first in Mathematics and Physical Science and he continued with research work at the College for a short time. In 1894, he was awarded a scholarship which had been endowed at the time of the 1851 Crystal Palace Exhibition. This enabled him to go to Trinity College, Cambridge, as a research student at the Cavendish Laboratory. In 1898 he identified alpha and beta rays in uranium radiation and indicated some of their properties. In the same year he left for Canada to take up the Chair of Physics at McGill University, Montreal. There he discovered a new gas, thoron, an isotope of radon. With Soddy he created the "disintegration theory" showing that radioactivity is an atomic and not a molecular processes. Otto Hahn, who later discovered atomic fission, worked under Rutherford in 1905–06.

In 1908 Rutherford was awarded the Nobel Prize for Chemistry for his investigations into the disintegration of the elements and the chemistry of radioactive substances. By then Rutherford had returned to England as Professor of Physics in the University of Manchester and, working with H. Geiger, he invented a method of counting the number of alpha particles emitted from radium.

In 1910, came his greatest contribution to physics when scattering of alpha rays showed the "nucleus" containing almost all the mass of the atom to be concentrated in a minute space at its center. In 1912 Niels Bohr joined him at Manchester. Bohr adapted Rutherford's nuclear structure to Max Planck's quantum theory and so obtained a theory of atomic structure which, with later improvements, mainly as a result of Schrödinger's and Heisenberg's concepts, remains valid to this day; i.e., that the nucleus is at the center of an atom, surrounded mostly by empty space, and carrying most of the weight of an atom.

In 1913, with H.G. Moseley, he used cathode rays to bombard atoms and showed that their inner structure corresponds with a group of lines and an atomic number which characterizes the elements and defines its properties. In 1919, during his last year at Manchester, he discovered that the nuclei of certain light elements, such as nitrogen, could be "disintegrated" by the impact of energetic alpha particles coming from a radioactive source, and that during this process fast protons were emitted. Blackett later proved that the nitrogen in this process was actually transformed into an oxygen isotope, showing that Rutherford had been the first to deliberately transmute one element into another.

From Manchester he returned to the Cavendish Laboratory where he was an inspiring leader steering many Nobel Prize winners towards their great achievements including Chadwick, Blackett, Cockcroft and Walton. Rutherford's chief recreations were golf and motoring.

He was buried in the nave of Westminster Abbey, near Sir Isaac Newton and Lord Kelvin.

---

challenge the reader to find a common thread: Perhaps early interest in things scientific, perhaps the influence of role models, perhaps simply a driving desire to learn about the world, perhaps something else.

We very much expect that our story will appeal to students of many ages, in high school, college, graduate school and beyond. We hope that some will be inspired by this story and decide to become a part of it. Even if there were to be just a few, it would more than reward our efforts.

The physicist among our readers will discover that these "machines" hinge upon relativistic dynamics, electrodynamics, and plasma physics and, in their less disciplined moments, the chaotic behavior of non-linear systems. The engineer will find they embody a wide spectrum of advanced technology, from the production of high vacuum and the generation of large quantities of radio frequency power to the construction of superconducting magnets and superconducting radio fre-

quency cavities. Computer enthusiasts will be interested in the control systems, needed to ensure that everything runs smoothly, and the world's most powerful "number crunching" computers, which analyze the results from the massive particle detectors — today's equivalent of Rutherford's fluorescent screen. It is no accident that essential features of the Internet were the brainchild of accelerator laboratories and that its successor, the Data Grid, comes from the same stable. We shall describe all this and try to throw some light with technical sidebars on the more arcane technical jargon as we go along, but it is the machines themselves, and the scientists who made them possible, that are the main subjects of this book.

However fascinating the machines themselves, we should not lose sight of the purpose of accelerators that has sustained their development over eight decades and still urges us to plan even more ambitious machines for years to come. They are tools of particle physics — huge microscopes to help us see the smallest constituents of matter and understand the basic forces governing their behavior — a quest that started in the late nineteenth century with the unification of electricity and magnetism. It continued in the twentieth century as quantum physics, creating order out of the phenomenology of chemistry, and went on to explain the structure of the nucleus itself.

Since Rutherford's studies of the atom, particle physicists have studied smaller and smaller objects. Much later in the book, Figure 6.14 shows how the "fundamental" particles, protons, neutrons and electrons, relate to atoms and to their own smaller constituents: quarks. Experimental techniques for observing these particles have developed hand in hand with the accelerators themselves (Chapter VII).

Nearly all particle physics experiments, like those of Rutherford, are done by scattering. The technique is rather like trying to visualize the shape of a motor vehicle in the dark by kicking a soccer ball at it and recording the path of the deflected ball. Although such an experiment has to our knowledge never been performed by serious and mature scientists, it would be perfectly possible. These days one might use a computer to digitize and analyze the pictures of the ball's path taken by two video cameras. Imagine that the direction of the ball could be traced back to the point of impact; then the spread in these points would indicate the size of the object and perhaps whether it was a lorry or a car. Having identified a point of impact, the inclination of the car's bodywork at that point might be deduced from the angle of deflection. From this the contours of the body would emerge. The accuracy or resolution of detail would depend on the size of the football — a tennis ball would be an improvement — but you would still not be able to identify details like model number whose dimensions are smaller than the projectile. To resolve smaller details, you would need even smaller projectiles, such as dried peas or lead shot.

About the time Rutherford was experimenting in his hutch, physicists began to realize that particles such as alphas were not just hard spheres but can be also thought of as waves. De Broglie was able to show that the effective wavelength of these waves was inversely proportional to the particle's momentum. Waves, like those breaking on a shoreline, flow around objects that are smaller than their wavelength but tend to be deflected by objects, such as a breakwater, that are as large as, or larger than, their wavelength. Take the example of a 10 MeV alpha particle from radioactive decay of radium, as used by Rutherford in his experiments. Its wavelength is about 4×10^{-13} cm and about the size of the nucleus of the atom. Rutherford's scattering experiments were like kicking a football at another football in the centre of an arena. Once in a while, the alphas were deflected from the nucleus, but most of the atom proved to be empty space — the football would be undeflected, showing just how small the core of the atom is.

To see inside the nucleus, and reveal the structure of the protons and neutrons within, would need small projectiles corresponding to even shorter waves. Alpha particles would do; so incidentally would protons or electrons. According to de Broglie's law the wavelength is inversely proportional to the energy. Thus we see the ongoing need for higher energy particles and ever bigger accelerators.

Looking at this in another way, it is no coincidence that the alphas from radioactive decay of a nucleus are about 10 MeV in energy. This is a measure of the energy with which the strong force binds each of the particles in the nucleus together. Higher energy projectiles are needed to break the nucleus apart and even higher energy to prise the quarks from inside protons and neutrons.

The other benefit of using higher momentum projectiles is that their energy may be used to create new particles in the melting pot of a collision with a target particle. To accelerate a particle to 10 MeV between two metal plates, one would have to apply a voltage between them of ten million volts — 10 mega-volts. The energy is related to the mass of the desired product by Einstein's famous $E = mc^2$. To create a pion or other meson the energy has to be a few hundred MeV. At least 1000 MeV (1 GeV) is needed for a proton or antiproton; and for a Higgs particle, more than 100 GeV. The discovery of one of these heavy particles that would provide the missing part of the jigsaw puzzle of theories proposed to explain the structure and forces between particles. The Higgs, for instance, would for the first time explain why the electron is 2000 times lighter than the proton, though we must admit that such an explanation is too mathematical for the pages of this book.

Particle physics today makes a major contribution to our understanding not only of the submicroscopic nature of our universe, but how it was formed. Astrophysics and cosmology draw upon the most recent discoveries at accelerator laboratories to model the interplay of forces and particles at the moment of the Big Bang. Such models can predict the amounts of light elements (helium, deuterium and lithium) in the early universe. The velocity of sound in these early moments may be calculated, and consequently the cosmic microwave background — still with us today; in fact, both its magnitude and its variations can be estimated. Taken together with astrophysical observations of supernovae and galaxy clusters, modern physics finds that the universe must contain "dark matter" and "dark energy" and these together must be its major constituents — far outweighing normal and observable matter. Since this dark matter cannot be seen by astronomers, we must seek its explanation in the host of new, supersymmetric particles that are the conjecture of particle physics beyond the standard model. At the moment such theories pose more questions than they answer, but should supersymmetric particles exist, users of the accelerators and colliders of today and tomorrow will strive to reveal their secrets and with them, the ultimate secrets of the universe from quarks to galaxies.

The chapters of this book follow a historical pattern. In the early years a generation of electrostatic machines was developed (Chapter I) to mimic the energies from natural radioactivity. Then, in the 1930s, cyclotrons were invented to accelerate protons and deuterons to energies ten times larger, but the energy of these machines was limited by their sheer bulk (Chapter II). Linear accelerators were also developed for accelerating both protons and electrons (Chapter III). Efficient electron acceleration required a different kind of device — the betatron (Chapter IV).

Soon after accelerator builders returned to peaceful work after World War II came the two breakthroughs that opened the door to acceleration to almost unlimited momenta — the synchrotron (Chapter V) and the collider (Chapter VI). Since then, synchrotrons and colliders of higher and higher energy have been built to reveal finer and finer detail.

The first of these major developments was the invention of the synchrotron, which experimenters used to create new particles by firing a high-energy particle at a stationary target. As energies climbed, this process became less and less efficient. This is because in a collision with a stationary object only the square root of the kinetic energy of the projectile is available for making new matter; the rest is "wasted" to provide the kinetic energy of the outgoing products of the collision as they stream forward following the impact. It was realized that if two particles collide head-on, all of their energy is available for particle production. The second major development, the collider, came as a result of developments in accelerator science that allowed physicists to collide one beam against another. Colliders could store beams rotating in opposite directions and collide them in the middle of experimental apparatus — apparatus

that grew into huge assemblies of particle detection equipment (Chapter VII).

Thanks to Rutherford and his legacy of the understanding of nuclear fission, most of the resources and effort in the accelerator field have, for a century, been directed first towards nuclear and then sub-nuclear physics. In the period since World War II, the focused has moved to sub-nucleon physics. Nuclear physics (sub-nuclear) is concerned with nuclear structure and how a nucleus "works." It has brought us the atomic bomb, nuclear power reactors, and nuclear medicine. High-energy physics — sub-nucleon physics — is concerned with the particles that make up the nucleus, i.e., quarks, and with particles that must be created at high energy and only live a tiny fraction of a second, such as mesons. It is also concerned with neutrinos.

Pursuing sub-nucleon physics to higher and higher energies has been the main driving force behind accelerator development and has led to a few large and expensive facilities, supporting thousands of university and laboratory research workers.

The study of sub-nucleon physics is certainly in the great tradition of seeking knowledge for its own sake and may some day be of practical importance to mankind, but to date has proved to be of little value or relevance to the man in the street. Indeed, the man in the street may be forgiven for believing that the number of families of neutrinos seems about as important to mankind as the ancient theologians' arguments to establish how many angels are able to dance on the head of a pin! A scientist would of course argue that the number of neutrinos is not an arbitrary number, but something that is gleaned from nature with great effort and is an eternal truth worth pursuing for its own sake in our struggle to understand the world around us — quite different from the activities of theologians.

Nevertheless, perhaps one may be glad that the trend in recent years is for the governments to put even more money into accelerators that serve other applications than particle physics, with the result that now there is an even larger community of researchers served by accelerators in other ways.

Although the benefit to mankind of pure research cannot be measured in money it is important to realize that the investment in pure research, which initially supported the development of particle accelerators, has been returned many times over in the benefits from later application of these machines to other fields and new technologies. We shall touch on some of the applications and highlight two major applications: namely synchrotron radiation and cancer therapy.

Turning first to synchrotron radiation, recent years have seen the construction of dozens of electron synchrotrons of a few GeV energy. Their purpose is to generate the ultraviolet and x-rays that stream out from a beam of electrons when bent into a non-linear (for example, a circular) path. The scattering of this synchrotron light from a complicated protein or enzyme molecule can reveal its structure in astonishing detail. This is the key to many of today's life sciences questions, and can equally well reveal the structure of the surface layers and crystal materials that are the basis of semiconductor development. Synchrotron radiation facilities are treated in detail in Chapter VIII.

The second application that we highlight, cancer therapy, is treated in Chapter IX. A recent census of accelerators of all energies found over 14,000 machines of all energies worldwide. Almost half of these are in the hospitals of the developed world and are used either to produce isotopes for medical imaging or, more often, for cancer therapy. We discuss the widespread use of linacs as x-ray sources for therapy, and also the use of hadrons (200 MeV protons and 400 MeV carbon), from cyclotrons and synchrotrons, for therapy. Hadron therapy is particularly useful for treating tumors located near sensitive organs.

Finally, in Chapter X we review the past and the present and attempt to foresee the future. We have seen that not everyone has followed the road map of Rutherford, and in fact the use of accelerators for applications other than high-energy particle physics dominates. In that Chapter we indicate some of the facilities of the future, which shall be used to continue, and advance, sub-nucleon studies. We also review some of the other practical applications of accelerators.

One such application, which involves perhaps half of the accelerators in the world, can be described as accelerators in an industrial setting. These are either low energy electron machines used to irradiate plastics and rubber to make them harder, or to sterilize food and grain; or ion accelerators to implant ions in, for example, the production of most high performance computer circuitry. Accelerators are also used for non-destructive testing and for screening the cargos of trains and trucks.

Another important application is in the generation of neutrons. Neutrons can be used, rather like synchrotron light, to reveal the structure of molecules and materials and although they cannot be accelerated they may be produced in copious numbers when an intense beam of (about) 1 GeV protons bombards a heavy metal target; this is called spallation.

Other applications, such as heavy-ion fusion, accelerator driven nuclear power reactors, and the accelerator treatment of nuclear waste are also described. We then look to the future and describe machines soon to come to pass; dreams of machines; and the very far future of accelerator science when lasers and plasmas will play a role.

Although the format of this book is roughly chronological, it is not intended as a textbook, but rather as a collection of material for the reader to browse through at leisure as one might read a "coffee-table book." Each chapter is more or less independent. Although we first explore the roots of accelerators and then follow each branch of the accelerator tree out to where it now stands, most chapters can be read and appreciated independently. We include a glossary of technical terms as Appendix C, pp. 175–183. We hope that readers will find that the mixture of historic photographs, biographical sketches and technical explanations will conjure up an understanding of some of the enthusiasm, frustration and successes of the principal contributors to this field.

Chapter I. Electrostatic Accelerators

I.1 Scientific Motivation

It was Rutherford who fired the pistol to start the race to build accelerators of ever increasing energy. By 1928 his reputation as a founding father of nuclear physics was firmly established and, as a reward, had just been made president of the Royal Society of England. In his presidential address he said, "I have long hoped for a source of positive particles more energetic than those emitted from natural radioactive substances." These were inspiring words and no doubt they focused the attention of many new minds as well as encouraging others who were already working on the problem.

At that time, the cyclotron had yet to be invented — though that was to be only a year or two later. Two acceleration methods had been proposed: the linear accelerator by Ising and the ray transformer of Wideroe. We shall devote later chapters to these machines, but when Rutherford made his plea they would have to wait many years before they became practical. Meanwhile many nuclear physicists pinned their faith in the electrostatic route — the simplest concept of all — acceleration between the terminals of an electrostatic high voltage generator. Notable among these were Cockcroft and Walton at the Cavendish Laboratory in Cambridge.

One of the first applications of electrostatic accelerators at the Cavendish was to fire particles into the nucleus. It had been calculated from classical mechanics that a particle would have to have a very high energy to penetrate the nucleus. Rutherford had used 10 MeV alpha particles from radioactive decay for his famous scattering experiments and it was argued that if the nucleus can emit particles with 10 MeV energy, the forces which bind it together must be strong enough to repel any particle approaching it with a smaller energy. Ideas for suitable high voltage generators were very much in their infancy and it was impossible to imagine matching the 10 MeV energies of alpha particles. Even if someone had succeeded in making a 10 MeV power

supply the problems of avoiding catastrophic breakdown to ground would have been challenging. Sparks can jump meters at that voltage. There was a long way to go.

Fig. 1.1 The Cavendish Laboratory, where the landmark experiments of Cockcroft and Walton took place. Note the ecclesiastic style of university buildings of that time. This laboratory was the setting for the many wonderful innovations and developments described in the sidebar.

1

Sir John Douglas Cockcroft (1897–1967)

English physicist
Nobel Prize for Physics, 1951 (shared with Walton)
Co-inventor of the Cockcroft-Walton high voltage generator

John Cockcroft was the eldest of five sons of a Yorkshire miller. He won a scholarship to study mathematics at Manchester University but became fascinated by atomic physics after reading the work of J.J. Thompson and Rutherford — then at Manchester. His studies were interrupted by the First World War, in which he served as a junior officer, and was the only survivor of his unit during bitter fighting at a forward post on the Western Front.

After completing a Manchester degree in electrical engineering, he joined the Metropolitan Vickers Company but was eventually sent to Cambridge to improve his mathematics. There, he again came across Rutherford, who promised to take him into the Cavendish if he got a "first" in mathematics. Meanwhile he was to be "spare time honorary electrical engineer." He first worked with the famous Russian, Kapitza, who was intent on producing intense magnetic fields by short circuiting a very large generator through a single turn of copper. Cockcroft was able to draw on the experience of his Metropolitan Vickers friends to help. Kapitza's bold use of electrical machinery perhaps inspired Cockcroft to take the direct approach when the time came for him to accelerate protons. Together with Walton he invented the high voltage generator that bears their name. There were shades of Rutherford's gift for improvisation when they scoured the countryside for the discarded glass sight glasses from petrol pumps, which they stacked up to form the acceleration tube.

Cockcroft and Walton used this apparatus in 1932 for their momentous experiment: the first artificial transmutation of lithium into helium, which was to much later gain them the Nobel Prize.

Unlike his retiring partner, Cockcroft continued at the Cavendish until, like Oliphant, his skills were redirected at the start of the Second World War towards securing Britain's air defenses. He was then sent to Canada to become director of the Montreal and Chalk River Laboratories.

His post-war career continued when, as Director of the Harwell Atomic Energy Research Establishment, he led the development of nuclear energy for peaceful and defense purposes. He was much respected by his 6,000 or so staff, many of whom, however junior, would have a tale to tell of how he had wandered unannounced into their workshop or laboratory to show an interest in what they were doing. After a few encouraging words, he would make miniscule crisp notes in a famous little black book about what he should remember and attend to in order to help their efforts.

Ernest T.S. Walton (1903–1995)

Irish physicist
Nobel Prize for Physics, 1951 (shared with Cockcroft)
Co-inventor of the Cockcroft-Walton high voltage generator

Ernest Thomas Sinton Walton was born at Dungarvan, County Waterford on the south coast of Ireland on October 6th, 1903, the son of a Methodist minister from County Tipperary. In 1915 he was sent as a boarder to the Methodist College, Belfast, where he excelled in mathematics and science. In 1922 he entered Trinity College Dublin and graduated in 1926.

In 1927, he went to Cambridge University to work in the Cavendish Laboratory under Lord Rutherford, receiving his PhD in 1931. At the Cavendish Laboratory, he worked on methods for producing fast particles, first on a linear accelerator and then what was later to become known as the betatron. He followed this with work done jointly with J.D. Cockcroft on the direct method of producing fast particles by the use of high voltages. Together they invented a column of diodes, known to this day as the Cockcroft-Walton generator, which could multiply a modest voltage into several hundred thousand volts. Fast protons were accelerated down an evacuated glass column between the terminals of this device. The protons were used to bombard and disintegrate the nucleus of the lithium atom. The products were identified as helium nuclei.

This was the first occasion on which an atomic nucleus of one element had been artificially changed into that of another element — the so-called "splitting of the atom." This discovery, for which he was later to receive the Nobel Prize, was arguably the most important and momentous made by Rutherford's brilliant team of researchers at the Cavendish Laboratory in the mid 1930s.

A reserved and retiring man, Walton was soon to leave Cambridge. In 1934 he returned to Trinity College, Dublin, as Fellow and then Professor of Physics.

But as quantum mechanics began to be understood, George Gamow realized that the uncertainty principle allowed particles to escape from nuclei at a lower energy and that this would explain radioactive decay. E. Condon and R.W. Gurney had independently come to the same conclusion. Gamow then turned the argument around and predicted that quantum mechanics would allow nuclear particles to enter the nucleus at significantly lower energies than had been thought. He estimated that 300 keV might be enough. (See box on the Coulomb Barrier.) Gamow's work became the guiding light for John Cockcroft and Ernest Walton of Cambridge, who set about building an electrostatic generator for this voltage.

The Cavendish Laboratory in the Time of Cockcroft and Walton

In a narrow lane opposite Corpus Christi College, Cambridge, its site chosen to protect it from the vibration of traffic, stands the Cavendish Laboratory. It was established in 1871 and its first professor was the great James Clerk Maxwell who, being a great admirer of Cavendish's work on electricity, christened the laboratory after him. From the outset it was to be a laboratory in which the highest standards of experimental physics were to be observed using the simplest of apparatus — often hand built by the researcher himself.

Accelerators began to be of importance at the Cavendish Laboratory in 1927. Nobel prize-winner Ernest Rutherford was then its director, following in the footsteps of Maxwell, Lord Rayleigh and J.J. Thomson, each of whom had made revolutionary contributions to our understanding of physics.

One might not expect to find the atmosphere among such eminent physicists to be light hearted, yet the Cavendish at the end of the 20s could be quite informal. The working day hardly started before ten in the morning, and every evening at six, a technician would solemnly go the rounds to switch off all lights and power supplies, even if this meant that a crucial measurement had to be interrupted and repeated the next day. A visiting American is reported to have written to Thomson that by US standards this would be seen as "sloth and indolence" — but that was the style; and Thomson held that experiment should always be followed by a respectable period of reflection.

The frequent colloquia and seminars at the Cavendish were of the highest standard and invited speakers came from all over the international world of physics. In Rutherford's time there were as many as fifty graduates studying for PhD's and it is not surprising that there was also some fun to be had. The more senior researchers were not above joining in, and Peter Kapitza, then working at the Cavendish, describes the lab's annual dinner where: "you could do anything you liked at table — squeal, yell, and so on. After the toasts it was very funny to see such famous luminaries as Thomson and Rutherford standing on their chairs and singing at the top of their voices."

For Rutherford all science, indeed all physics other than the search for atomic structure, was "mere stamp-collecting." He was famous for employing only the simplest of apparatus. It was perhaps then a surprise that he pressed for expensive electrical equipment to provide a steady stream of charged particles for research at the Cavendish, but his research seemed to have reached an impasse. It was known then how strong the forces confining the positive charges of protons in the nucleus must be and it seemed clear that many MeV of energy would be needed for a particle to break in and probe the structure of the nucleus. After all, the particles emitted from nuclear decay have energies of this magnitude. Electrical industry did not seem capable of producing steady voltage of this size and attempts by Walton to build betatrons and linacs had not been successful. In the end, the energy needed did not turn out to be that high and was just within the reach of electrostatic machines. The way in which Rutherford came to be convinced of this is interesting and says much for the effective international communication of ideas at the time as well as Rutherford's open mind.

In the years after the First World War experimental research into the structure of the atom was centred in Germany, France and the United Kingdom. Rutherford, first at Manchester, then at the Cavendish, was the focal point for the majority of the experimental work while Bohr's institute in Copenhagen headed the theoretical work, hosting the fathers of quantum mechanics — Heisenberg, Schrödinger and Pauli. Copenhagen was not unlike the Cavendish in its informality and lively social life. There was a healthy cross-fertilisation between the two centres and it was thanks to a friendship between Mott of Cambridge and Gamow, a Ukrainian physicist at Copenhagen, that the idea of applying wave mechanics to nuclear disintegration reached Cambridge.

Gamow inherited socially colourful and rather bohemian characteristics from his Odessa family. His father had taught the adolescent Trotsky, and Gamow junior had taught Red Army artillerymen the basics of physics. His bedroom wall boasted a framed verse "when the morning rises red, it is best to lie in bed." He nevertheless was both brilliant and productive physicist and soon moved on to Bohr's Institute by way of Göttingen. His contribution to the work of Rutherford's Cavendish was however crucial.

It had seemed impossible to explain how the alpha particles that emerged from the nucleus in radioactive decay could get out against the strong forces binding the nucleus together, and this had been a major stumbling block in understanding the atom and its structure. Gamow showed that the uncertainty principle and the wave nature of particles would allow a charged particle a small probability to enter or leave at an energy of 300 keV: much smaller than had been previously thought necessary (see sidebar on the Coulomb Barrier).

Mott and Hartree from the Cavendish visited Copenhagen and, although with hardly a word in common, were thrilled with Gamow's tales of the Russian Revolution. They were even more interested in his theory suggesting the nuclear barrier might be penetrated and encouraged him to bring it with him to the Cavendish.

Rutherford was naturally wary of new-fangled theories such as quantum theory and wave mechanics, but Gamow convinced Rutherford of his ideas during a visit to the Cavendish. When Cockcroft, then supervising the work of Walton and Allibone, was asked for five hundred pounds for an electrostatic generator to produce the 300 keV that had Gamow calculated, he readily agreed.

Allibone, like Cockcroft, had previously worked at Metro Vickers and was no stranger to electrical machinery, while Walton had been already trying to build other accelerators. The work of constructing the high voltage generator started and, when concluded, made it possible to induce a series of nuclear reactions with accelerated particles. This brought experimental physics at the Cavendish in the 1930s out of the era of string, sealing wax and glass blowing. Research directors had to be convinced, financial approval sought and the professional help of consulting engineers engaged. Of course similar changes in the approach to experiments were also underway in Lawrence's laboratory in California.

At the same time, a previous colleague of Rutherford from Manchester, Hans Geiger, had invented an electronic tube to count particles. This, championed by Chadwick at the Cavendish, together with Cockcroft and Walton's electrostatic accelerator, brought a whole new dimension to the speed and accuracy of observation.

After pioneering atomic structure research, the Cavendish went on to produce four more Nobel laureates including prizes for unravelling the structure of DNA. It remains at the cutting edge of research in the UK.

I.2 Voltage Multiplying Columns

Cockcroft and Walton's accelerator was based upon the idea of a voltage multiplying column (Fig. 1.2). Normal alternating current was first rectified (converted to direct current) and then applied to a number of large condensers (capacitors). The condensers, each bearing a voltage of a few hundred volts, were then connected in series so the voltage of each one added to the next. The combined voltage, in this case 600 kV, was then connected to an accelerating column in order to accelerate protons.

Their first publication, in 1932, was followed a few months later by another describing the very first man-made nuclear reaction. In this, a proton was added to a lithium atom to produce Be-8 that decays to 2 helium nuclei. In fact this crucial experiment used their second accelerating column. They had treated Gamow's figure of 300 keV with more caution than it deserved and had become fascinated with the task of going to 600 keV or more. They were bent upon improving their machine even further before making the measurements but Rutherford, the head of the Cavendish at the time, stepped in and insisted that they try the experiment without waiting for these improvements. In fact, even their first machine would have been adequate, for Gamow had been very correct in his estimates! Cockcroft and Walton were honored, much later, by receiving the Nobel Prize in 1951.

In the years that followed many devices were constructed, based upon the concept of the voltage multiplying column invented by Cockcroft and Walton. A 600 kV system was built in Ottawa in the late 30s and commercial systems were built by the Phillips Corporation before and after World War II. In the late 1960s a 2 MV system was installed at the Cavendish Laboratory and a pre-injector of 750 kV was built, and used, at the National Accelerator Laboratory (now Fermilab) in Batavia, Illinois (Fig. 1.3). This 2 MV column is about as far as one can go with a device in the open air before spark breakdown occurs. Enclosed systems, employing high pressure insulating gases, have achieved 6 MV. There was always a need for even higher energy and improvement in reliability and ease of operation.

Fig. 1.2 The original Cockcroft-Walton installation at the Cavendish Laboratory in Cambridge. Walton is sitting in the observation cubicle (experimental area) immediately below the acceleration tube, which was covered with black velvet so that the faint scintillations might be observed by the detector.

Fig. 1.3 The Cockcroft-Walton pre-accelerator, built in the late 1960s, at the National Accelerator Laboratory in Batavia, Illinois. This very large and expensive installation provided the voltage for the first tiny step in the acceleration of protons to energies of hundreds of GeV.

I.3 Silk Belts

Even before Cockcroft and Walton, Robert Van de Graaff, then a mechanical engineer studying physics at the Sorbonne, was inspired by a lecture given in 1924 by the very famous Louis de Broglie. Like Rutherford, De Broglie hoped for a machine to allow the careful and controlled study of nuclear particles and van de Graaff immediately set himself the task of building such a machine. From 1925–1929 he was at Oxford (on a Rhodes scholarship) and during this time he developed the concept that, ever since, has been called the Van de Graaff generator. (See sidebar for Van de Graaff and Fig. 1.4.)

In 1929 he went to Princeton, on a National Research Council Fellowship, and during this year built the first working model (only 80 kV). His idea was that the machine would transport charge mechanically (all other accelerators, before and since, have been purely electrical). He sprayed charge on a moving belt of pure silk which transported it to an upper terminal where it was removed. The device built up charge and hence

Robert Jemison Van de Graaff (1901–1967)

American physicist
Inventor of a high voltage generator used as an accelerator

Robert Van de Graaff was born in Tuscaloosa, Alabama to Minnie Cherokee Hargrove and Adrian Sebastian Van de Graaff. He followed a degree course in mechanical engineering before starting work in the research department of the Alabama Power Company. It seems that Alabama was not at that time crowded with young men wanting to pursue their studies abroad and he was fortunate enough to get a place to study at the Sorbonne in Paris from 1924 to 1925, where he attended lectures by Marie Curie on radiation. In 1925 he went to Oxford University in England as a Rhodes Scholar, where he received a PhD in physics. It was there that he became aware of the plea of Ernest Rutherford that particles might someday be reliably accelerated to speeds sufficient to disintegrate nuclei.

Like Wideroe (see sidebar in Chapter III) he was inspired to respond to this challenge, but in quite a different way. In his physics classes, Van de Graaff would certainly have seen electrostatic machines, in which charge, sprayed on a metallic segment, could be transported and extracted onto a high voltage terminal. He extended this idea to use a moving belt to charge up a sphere. An ion source, within the sphere, generated particles, which were then accelerated as they traveled down to ground potential through an evacuated column.

In 1929 he returned to the United States to join the Palmer Physics Laboratory at Princeton and constructed the first working model of an electrostatic accelerator, which developed 80,000 volts. Improvements were made to the basic design and in 1931, at the inaugural dinner of the American Institute of Physics, a demonstration model was exhibited that produced over 1,000,000 volts.

It was later realized by Bennet in 1937 and, even later, by Alvarez in 1951, that by using a tandem Van de Graaff to accelerate negatively-charged ions to a high voltage and then stripping them of their charges with a foil, they would return to earth potential with twice the voltage of the generator.

At the time of his death there were over 500 Van de Graaff particle accelerators in use in more than 30 countries.

a voltage on the terminal which was in the form of a large sphere: the shape that best holds charge without creating a spark. This voltage was then applied (as in Cockcroft-Walton machines) to an accelerating column down which particles accelerated from high voltage to ground. In later higher voltage versions of the machine, the belt was housed inside an insulating gas, sulfur hexafluoride, to minimize sparking.

Van de Graff's first published device achieved 1.5 MeV. This was in 1931 and actually before Cockcroft

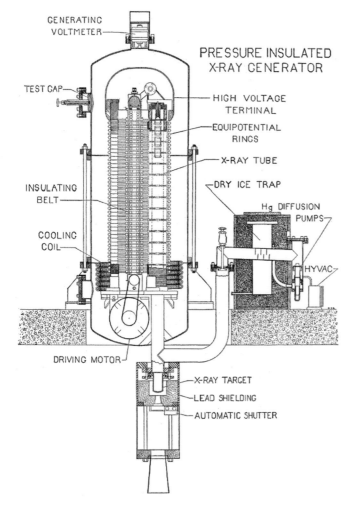

GENERATING VOLTMETER

PRESSURE INSULATED
X-RAY GENERATOR

TEST GAP

HIGH VOLTAGE TERMINAL

EQUIPOTENTIAL RINGS

X-RAY TUBE

DRY ICE TRAP

Hg DIFFUSION PUMPS

INSULATING BELT

COOLING COIL

HYVAC

DRIVING MOTOR

X-RAY TARGET
LEAD SHIELDING
AUTOMATIC SHUTTER

Fig. 1.4 Line drawing of a Van de Graaff accelerator — part of one of Van de Graaff's patent applications around 1930.

Odd Dahl (1899–1994)

Norwegian physicist
Builder of three Van de Graaffs, a betatron and a nuclear reactor in Norway
Led the design of the first CERN Proton Synchrotron

Odd Dahl was one of the disproportionate number of Norwegians who pioneered novel and successful accelerator projects. With only a modest formal education he joined Amundsen's 1922 Arctic expedition as an air pilot. After his plane was damaged beyond repair during a difficult takeoff from an ice floe, he spent the following two ship-bound years learning physics and developing oceanographic instruments. In 1926 he joined the Carnegie Institution in Washington where, working with Merle Tuve and Lawrence Hafstad, he developed instruments for the study of terrestrial magnetism of the atmospheric Kennedy-Heaviside layer and for nuclear physics.

Returning home to the Christian Michelsen Institute in Bergen in 1935, he built a new series of three Van de Graaff machines, a betatron and, in the early days after World War II constructed Norway's first reactor without access to classified work.

In the very early days of CERN (1951) he was appointed to build their first 10 GeV proton synchrotron but, after hearing of the discovery of alternating gradient focusing during a historic meeting at Brookhaven with its inventors, Courant, Livingston and Snyder, he returned to tell the CERN team that "you must drop everything and work only on this."

Thanks to his courage and leadership, CERN physicists were able to switch to a machine of three times the energy, bringing Europe on a par with events in the USA. He was mentor to Kjell Johnsen (see sidebar in Chapter VI on Johnsen), another Norwegian whom he recruited in the early days of CERN and who went on to build the ISR project. Johnsen said of him "he accepted only challenging tasks and his intuition never failed him" — a comment that might equally apply to Johnsen himself and to another Norwegian, Bjorn Wiik (see sidebar in Chapter VI on Wiik) who was later to lead the HERA project at DESY.

and Walton completed their work. It was a dual device, 2 m tall that had two 5.6 cm silk belts feeding two spherical upper terminals. We are told it only cost $100 and it was patented a few months later.

The concept was immediately picked up by Gregory Breit, Odd Dahl, and Merle Tuve at the Department of Terrestrial Magnetism of the Carnegie Institute in Washington, who made many technical contributions that greatly improved the machines. (See sidebar for Dahl.) Soon they had constructed a device for a voltage of 1.2 MeV which, unlike Van de Graaff's, had a single two-meter upper sphere. It was assembled in the open air and sparked whenever an insect alighted on the sphere. Typical operating voltage, including the effect of insects, was only 600 kV. They later built a much more reliable device indoors which also had a two-meter upper terminal and produced 1.2 MeV.

Van de Graaff moved to MIT in 1931 and built a very large generator in a former balloon hangar (Fig. 1.5). The upper terminals were 4.6 m in diameter, 6.7 m high. Paper was used for the belt rather than silk and the voltage between the two terminals was 5.1 MV. It was difficult to maintain smooth non-sparking spheres because of pigeons living in the roof of the hangar. The effect of their droppings prompted some very dramatic pictures (Fig. 1.6).

I.4 Wisconsin Advances

About this time (1932) development of electrostatic accelerators also started at the University of Wisconsin and was pioneered by G.G. Havens. It was here that some improvements were built in to avoid the fickle

THE GENERATOR IN THE HANGAR AT ROUND HILL

Fig. 1.5 Van de Graaff's very large accelerator built at MIT's Round Hill Experiment Station in the early 1930s. The spheres stood 43 feet above the ground, supported on steel trucks that ran on a railroad track to make it possible to change the striking distance.

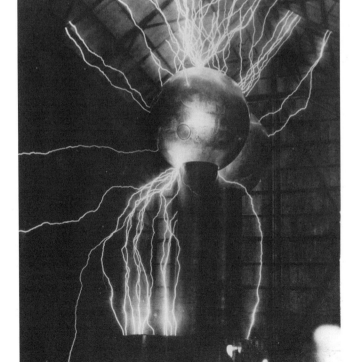

Fig. 1.6 Because their electrodes were very smooth and almost perfect spheres, Van de Graaff generators did not normally spark. However, the installation at Round Hill was in an open-air hangar, frequented by pigeons, and here we see the effect of pigeon droppings.

AT ROUND HILL SPARKING TO HANGAR (LONG EXPOSURE)

nature of the earlier machines. The first of these machines had not reached a high enough voltage to induce nuclear reactions, but by 1933, Havens and his students (including Donald Kerst, who invented the betatron, and Raymond Herb, about whom we shall say much more) had achieved 1 MeV and reproduced the nuclear reaction that Cockcroft and Walton had observed earlier.

In 1935, Raymond Herb, who had just returned from a summer at the Department of Terrestrial Magnetism, introduced the first of many technological advances in electrostatic machines. He took the original insulating vacuum tank of Van de Graaff and filled it with pressurized gases to improve the voltage holding capability of the charging belt and the accelerating column. He also put conducting rings along the accelerating column, which helped control the accelerating gradient. All modern machines use these two innovations. He went on to develop the mechanical engineering that allowed the whole machine to be cantilevered out to be operated horizontally. There were other technological advances too, which helped make the machines less expensive, easy to operate and highly reliable.

One of Ray Herb's most important inventions, first applied in 1965, was to change the charging belt from a continuous smooth belt to one with steel cylinders held apart by nylon insulating links in the form of a chain. These new drive systems were highly efficient, had a long life, needed little maintenance, provided extremely stable terminal voltage conditions and were in general far superior to the old belts. Herb called machines using these belts Pelletrons, and he was sufficiently enthusiastic and confident about the new machines to form a company to build them.

I.5 Tandems

It was W.H. Bennett, in 1935, who first realized that an electrostatic column would accelerate negative ions (neutral atoms with electrons added) from ground potential to the high voltage terminal — the opposite route to that taken by the positively charged ions that a Van de Graaff normally accelerated. If the electrons of the negative ions could be stripped of electrons by passing through a thin foil within the high voltage terminal, then the positively charged nuclei would be further accelerated as they returned down the column, thus apparently doubling the energy of a machine. In 1940, he patented the idea and in 1953 laid down in detail the design of a modern tandem accelerator. In 1951, Luis Alvarez, building on the work of the 30s,

Born on a small farm in Wisconsin, as one of eight children, he did his undergraduate and graduate work at the University of Wisconsin, where he received his PhD in 1935. He joined the faculty of the University of Wisconsin from 1935 until his retirement in 1972. Except for one summer in 1935, spent at the Department of Terrestrial Magnetism in Washington, and the war years of 1940–1945, at the Radiation Laboratory at MIT, he spent his entire life in Wisconsin.

Although Ray Herb produced some fine nuclear physics with the electrostatic machines he built, his special talent was in the technical field, making these devices ever better. It is perhaps difficult for the outsider to really appreciate his contributions. One discovery was that the dielectric properties of air could be increased by adding carbon tetrachloride so that one could get to higher voltages in electrostatic machines; another, that by adding voltage controlling rings, corona points or resistors all around the accelerating columns, one could get to even higher voltages; and there were many other "small things" that made his machines the best in the world.

We describe in this chapter how Ray Herb, in 1965, established the National Electrostatics Company. Although the start was difficult, the company ultimately employed up to 140 people and produced 130 machines. The "selling" of these machines, for purposes far beyond nuclear physics, was very much Ray Herb's specialty and after retirement from the University (at age 65), Herb devoted his full time to the company he had created.

Ray Herb was an enthusiastic leader, a talent he showed at an early age, during his time at the Radiation Laboratory during World War II and again at Wisconsin. He was an inspiring teacher and was cited by Eugene Wigner in his Nobel address as one of the three great influences in his life. Many others would agree.

In 1945 Ray Herb married Anne Williamson, a daughter of a physics professor, and they raised five children (one of whom became a physicist). He was an outdoors man and lived in homes with beautiful views. As civilization encroached, he moved twice, ever further from Madison, but always to rustic locations. He enjoyed watching wildlife from his homes and hiking along trails in the woods behind his houses. He was an enthusiastic canoeist, often spending weekends in that pursuit. Even at age 85, when in Sweden to receive an honorary degree, he took part in a lengthy canoe trip with his wife, his physicist son, and his son's wife. He combined the desire for recreation with very hard work and up until a few days before he died he was active full time in his work.

defined the necessary characteristics of the stripping foil and showed how to bend particles 180 degrees so that the same accelerating column could be re-used, cheating nature. He dubbed the device a "Swindletron." Nowadays such tandem machines are common.

I.6 Commercial Production of Electrostatic Machines

The specialized techniques needed to design modern accelerators are usually first developed in a large national laboratory or university department. Manufacturing industry is then asked to tender for the component parts a design with a rather precise specification, which the designers inspect during the manufacturing process. The result is that the manufacturers improve their expertise in certain fields of precision engineering, microwave technology or control systems and also their competitive technological position.

Logically the next step would be for the manufacturer to design and build complete accelerator systems, assuming the detailed design work and control of the manufacturing for itself. However, such are the technological leaps needed to keep pace with the growth in energy of accelerators for particle research that industry has no time to put to good use the techniques it has learned before it must learn more tricks to build a bigger and more advanced machine. The market for very large accelerators is also limited in number.

This was never the case in the manufacture of electrostatic machines; from early on specialist firms sold complete systems to many clients, and today industry offers complete cyclotrons for isotope production and hadron therapy as well as medium energy machines as synchrotron radiation sources on a turnkey basis.

In the case of electrostatic machines two companies were formed to produce accelerators on a commercial scale. One was the High Voltage Engineering Company (HVEC), formed in 1947 with John Trump as Chairman of the Board and Van de Graaff as Chief Scientist. The other was the National Electrostatics Corporation (NEC), formed in 1965 by Herb, J.A. Ferry and T. Pauly.

From 1947–1953 HVEC produced many generators operating up to 4 MV. These were both reliable and convenient. Building upon the experience developed at MIT, where a 12 MeV machine was constructed, the company went on to produce machines for as many as 15 different countries operating in the range from 6.0 MV to higher than 16 MV.

At the present time the company produces a large number of different electrostatic machines for a number

of different purposes such as ion implantation, ion beam analysis of materials, accelerator mass spectroscopy, etc.

The National Electrostatic Corporation, started in 1965, had a difficult beginning, as the demand for nuclear instruments was limited and no one wanted to risk a large amount of money on a company with no track record. However, finally a 25 MeV Pelletron was sold to the University of Sao Paulo and, once this was seen to work, another to Oak Ridge. Many orders then came in and the company has to date produced 160 stable and reliable Pelletrons, located in 38 different countries. The company still specializes in these machines which can range from 25 MV down to 1 MV units for ion implantation.

I.7 Applications of Electrostatic Machines

Anyone working in a large high energy physics laboratory may be surprised to learn that although the frontiers of accelerator technology have moved away from devices that accelerate particles to a few MeV, the use of electrostatic accelerators is widespread and ever-increasing. There are literally thousands of them, which are applied to a wide range of scientific and industrial tasks, far beyond the conception of their inventors.

The original application was of course nuclear physics research and many Van de Graaffs and Pelletrons have been used for this purpose. The fact that their voltage may be very precisely controlled allows a detailed study of nuclear reactions and in particular narrow resonances in energy. Their advantage over other types of machines for such precise work may be judged by an anecdote from the World War II project to build an atomic bomb at Los Alamos. Precise values of neutron nuclear cross sections were needed to design the bomb and for this purpose a cyclotron from Harvard, a Cockcroft-Walton from the University of Illinois, and two electrostatic machines from the University of Wisconsin were secretly shipped to Los Alamos. The majority of the data that was needed came from the electrostatic machines.

Another interesting application is in the extensive study of the reactions that take place in stars — rather low energy by nuclear physics standards, and therefore requiring exacting work. In 1983 a Nobel Prize was awarded to William Fowler, of Caltech, for his "theoretical and experimental studies of the nuclear reactions of importance in the formation of the chemical elements in the universe" — experimental work that was all carried out on electrostatic machines.

Nowadays cyclotrons (Chapter II) and linear accelerators (Chapter III) are used in medicine rather than electrostatic accelerators, but perhaps the very first use of a nuclear machine in a hospital was in 1937 when Van de Graaff and John Trump constructed an electrostatic accelerator operating in the 0.5 to 1.2 MeV range. This was installed in the Harvard Medical School where it produced x-rays for cancer therapy.

In recent years electrostatic accelerators have been put to some unusual purposes. At the University of California in Santa Barbara, one was used as a driver for a Free Electron Laser (which is a powerful source of coherent radiation). Others have been used to generate the beam of electrons employed in electron cooling of ions, protons, and anti-protons where there is a need to have a "cold" (i.e., very mono-energetic and well-directed) electron beam moving along with exactly the same velocity as the heavy particles that are to be cooled. The heavy particles intermingle with the electrons and transfer some of their transverse motion to the electrons, thus becoming, themselves, more mono-energetic and better directed ("cooled"). Meanwhile the electrons are "heated" as they are deflected and excited but they are constantly replaced with fresh, cold electrons.

Another use of electrostatic accelerators is in a technique for material analysis called proton induced x-ray emission or PIXIE. In this technique protons are used to excite the atoms of the sample so that they emit x-rays. By analyzing the spectrum of the x-rays one may identify the chemical elements present in the sample. The proton beam can be made very narrow so that, for example, in environmental studies, one can excite one tree ring at a time. From the chemical composition it is then possible to map the atmospheric conditions year by year throughout the history of the tree. PIXIE has also been employed at the Louvre and other museums to detect art forgeries. In another study in Finland it was used to study trace elements in oyster shells as well as in old lead glass. It is also used to study trace elements in mineral exploration for locating regions where diamonds are likely to be found and has been widely used for biochemical and biological investigations.

Electrostatic accelerators are used to accelerate ions in the production of semiconductors. These are usually simple voltage columns; not complex machines. The ion beams are a tool to implant atoms into the interior of a semi-conducting wafer in order to alter its electronic properties. The energy needed to do this is small but must be controlled with precision to determine the depth at which the ions are deposited in a multi-layer circuit. This is the basis for the large industry of chips, which are at the heart of computers, cell phones, watches, mobile music devices, most consumer home products, and even automobiles.

Still another use of electrostatic accelerators, where again just the voltage column is used, is for accelerating the beam of an electron microscope. These devices employ the very short wavelength of electrons — quantum mechanics tells us that this wavelength is much shorter than that of light and therefore requires very high energy. In this way one may "see" smaller objects than one can with the naked eye or with normal optical microscopes. Modern electron microscopes can see individual atoms, and this is useful for many purposes, such as, for example, the design of efficient catalysts.

Electrostatic accelerators are very much a part of our modern life, and they can show up in surprising places, such as inspecting cargo passing through the Channel tunnel connecting England and France.

Far from being a museum exhibit, the electrostatic accelerator has thus found many applications in modern technology. Indeed the majority of the world's accelerators are low energy electrostatic machines working away in industry or in chemistry, biology and materials science laboratories far removed from "big science."

Chapter II. Cyclotrons

II.1 The Anatomy of a Discovery

Most scientists yearn for that "Eureka" moment and are prepared to spend months, even years, worrying about a problem, hoping that eventually the final piece in the jigsaw puzzle will slot into place. For Ernest Lawrence, one of the founding fathers of the accelerator field, the moment came in the early spring of 1929 as he sat in the Berkeley University library reading a paper by Rolf Wideroe. Knowing little or no German, Lawrence was only able to understand the drawings and the maths — but that was enough to trigger his imagination (see sidebar for Lawrence). In the mind of the general public, discovery is seen as a fleeting moment in time when an inventive mind makes a great leap forward. But so often several of the ingredients of the discovery will have been prepared by other cooks before the chef arranges them neatly on the plate and adds the essential seasoning with a cry of, "Voila!" Lawrence's invention of the cyclotron was like that. His contribution was to find a practical way both to bend particles in a circle with a magnetic field and at the same time accelerate by applying a small incremental voltage. Neither idea by itself was original but he was the first to suggest combining the two of them in one device: the cyclotron. The idea had unexpected beauty because as protons are accelerated and spiral outwards in a constant magnetic field, their revolution frequency does not change. This is because the proton, as it gains energy, not only gains speed, but follows a larger circle; the two effects just cancel. The constant circulation frequency is important because the particles arrive always in time with the incremental accelerating voltage. The voltage is applied as the particle passes between two "D" shaped electrodes in the magnet gap, which together form a resonant cavity. The cavity is driven by a radio transmitter at this resonant frequency. If the magnetic field is correctly chosen the particle's circulation frequency will be in resonance with that of the oscillating fields in the resonator and the proton (or ions for that matter) can be accelerated in a continuous stream.

Other workers were on the same track as Lawrence. Denis Gabor, who in 1971 was awarded the Nobel Prize for his "invention and development of the holographic method," had been thinking along similar lines. Gabor had already suggested combining magnetic bending with radio frequency acceleration and Leo Szilard had actually patented the concept in 1929. The story of Max Steenbeck, a graduate student, who had had the idea as early as 1927, is even more tantalizing. He worked out the theory, but had been put off when told by one professor that the idea would be too costly and by a second professor that the idea would not work. In spite of this he persisted for a time and, on joining the Siemens Company, was urged to write up the idea and publish it. He was at first reluctant to do this, considering it to be "too simple" an idea, but finally did so only to be discouraged by a note from the editor requesting further information. With this final setback he gave up and withdrew the manuscript. There are many lessons in this story for would-be innovators! The reader may rightly wonder why Wideroe himself did not put the two ingredients together. After all he, too, had been motivated by Rutherford's stirring words at the Royal Society challenging someone to invent a steady and reliable electrical method of accelerating particles. Furthermore both ingredients — bending and accelerating with oscillating fields — appear in his publications at different times. The explanation is probably partly chance and partly that Wideroe had been discouraged from both these ideas. But to see how this happened, let us explore Lawrence's ingredients a little more.

Even before Wideroe and Lawrence — and probably from the turn of the century — the concept that a magnetic field would bend charged particles was well-known. In our chapter on betatrons we explain how Wideroe in his doctoral thesis of 1927 presented the idea of a betatron, a circular accelerator for electrons. This "ray-transformer" was taken up much later by Kerst with great success, who called it a betatron. Although it bent particles in a circle it was suited for electrons and would not work for the much heavier protons that Lawrence wanted to accelerate. This was because protons, which are almost two thousand times heavier than electrons, lumber along well below the velocity of light where Newton's classical laws still apply, while the electrons in a betatron travel close to the velocity of light, where increments of energy do not affect their speed. The ray-transformer had no accelerating electrodes or cavity but worked like a normal transformer in which the electron beam replaced the secondary winding. In the betatron, the field rises so that the electrons, all traveling with the same speed and very close to the velocity of light, remained at the same radius. To add RF accelerating electrodes to this to make it into a cyclotron did not occur to Wideroe. He was focused on accelerating electrons and it would not have worked for particles whose velocity was very much less than the velocity of light and varying as the particle gains energy. It was perhaps easier for Lawrence with no preconception to tie him down. Also, even though Wideroe's theory of the electron motion turned out later to be sound, his thesis advisor in Karlsruhe, Wolfgang Gaede, had told him it would surely fail and he should find another topic for his research — circular machines were no longer on his agenda.

The second idea occurs in Wideroe's thesis (the paper Lawrence was mulling over in the library) where he described quite a different experiment: the first step to develop a linear accelerator for ions. This accelerator was based upon an earlier, theoretical, suggestion by Gustav Ising. The idea was to apply voltages oscillating at a radio frequency (RF) to a series of "cavities" or "drift tubes" in such a way that particles, passing through them like a string, threading a row of beads, would be given a little extra energy as they passed each gap between the tubes. We shall discuss this idea in some detail in Chapter III for it is the basis of a linear accelerator or "linac" (see Fig. 3.1). The length of the tubes was chosen so that as particles gained speed they would arrive at each gap just when the oscillating electric field between the tubes was pointed in the direction to accelerate them. Thus at no time was there any large voltage and, yet, the ions could be accelerated to high energy. (See Chapter III.)

In following this idea Wideroe was set on accelerating in a straight line and again to take the imaginative leap of Lawrence to see that the same accumulation

Ernest Orlando Lawrence (1901–1958)

American physicist
Inventor of the cyclotron
Nobel Prize, 1938

Born in South Dakota, he did his undergraduate work at the State University, and then went to work on a Masters degree at the University of Minnesota under W.F.G. Swann. He followed Swann to the University of Chicago and then Yale, where he received his PhD. He arrived in Berkeley in 1928 and remained there for the rest of his life.

EO, as many called him, devoted himself, almost exclusively, after the invention of the cyclotron to building a laboratory consisting of scientists of diverse capabilities and interests. His development of a team to pursue goal-oriented science was, perhaps, his most important contribution. The Laboratory was a close knit group and the members were singularly devoted to him. He was a great optimist and usually his team was able to bring his dreams into reality, sometimes employing inventions not yet developed when EO first initiated a project.

From the earliest days he was an advocate of applying the cyclotron to medical purposes, even saving his mother's life with a novel, 1 megavolt x-ray tube that he and Sloan had developed. From 1940 on, EO primarily devoted himself to questions of national defense and, in particular, to the development of nuclear weapons. He was instrumental in developing the separated uranium used in the Hiroshima bomb but he was the only scientist in the country to approach Secretary of State James Byrnes (then the highest official in the US government with whom any scientist spoke) with a plea that the power of the bomb be demonstrated to the Japanese, not actually dropped on them. On the other hand he was a strong supporter of Edward Teller's quest for a hydrogen bomb, advocating the initiation of a second weapons laboratory, named Lawrence Livermore, at a site he had already been using for the Materials Testing Accelerator at the Naval Air Station in Livermore.

He played a very important role in developing the cold war arsenal of nuclear weapons, but in his last years was an advocate of arms control. This last activity severely affected his health and probably led to his early death.

of energy might be applied to a circular machine would have been too much to expect.

A discovery is not always immediately put into practice. Archimedes' cries of "Eureka" did not alert a crowd of plumbers eager to apply his discovery to the bathhouses of the ancient world. During the time before the inventor moves on to the next step and while still basking in his achievement, he will be very sensitive to any skeptical reception, as indeed we saw in the case of Wideroe. The next move must often await encouragement from a respected colleague. Although Lawrence had this idea in 1929, he didn't work on it at first (he was busy with other things). He was eventually encouraged, in early 1930, to pursue the idea by Otto Stern — the same Otto Stern who was later, in 1943, to be awarded the Nobel Prize "for his contributions to the development of the molecular ray method and his discovery of the magnetic moment of the proton." Lawrence had enough faith in Stern's judgment to put a former graduate student, Niels Edlefsen, on to the problem, but there is doubt that Edlefsen ever achieved acceleration before he left in the summer of 1930.

Another general observation we might make while on the subject of inventors and their inventions is that the inventive mind wants often to move on to other things, and the hard grind of making an idea work is left to graduate students with an abundance of time and youthful enthusiasm at their disposal. Lawrence was no exception.

In the autumn of 1930, he obtained two excellent graduate students: David Sloan and M. Stanley Livingston. Sloan was set to work mainly on Wideroe's linear device and Livingston on the circular concept. By the end of 1930, Livingston had achieved resonance, and acceleration for 41 turns up to an energy of 80 kV in a 4-inch cyclotron. Livingston had to overcome many practical obstacles, including the development of a source of protons and obtaining a vacuum in the beam

---

Milton Stanley Livingston (1905–1986)
American physicist
Constructed the first cyclotron
Co-invented alternate gradient focusing
Enrico Fermi Award, 1986
Member of the National Academy of Sciences, 1970

Stanley Livingston was born in Wisconsin. His father was a divinity student who soon became a minister of a local church. When Stanley was about five years old the family moved to southern California, where his father became a high school teacher, having found that a minister's salary was inadequate to support a growing family. On the side he bought a 10-acre orange grove and ranch.

Stanley grew up learning and doing all the chores on the ranch. He became acquainted and fascinated with tools and farm machinery; these skills became the foundation of his ability as a designer and builder of complicated scientific systems throughout the rest of his life.

He received an undergraduate education at Pomona College, a masters degree from Dartmouth, and a PhD from the University of California in Berkeley. There, under Lawrence's guidance he constructed, for his PhD, the first cyclotron. He used the 4-inch magnet previously used by Edlefsen, built a vacuum chamber and installed a hollow D-shaped accelerating electrode in it, added an RF oscillator, and put the whole system together. With Lawrence's active help and supervision he first found evidence of resonant acceleration about November 1930.

He went to work on designing and building the next machine, an accelerator designed to go to one million volts based on a magnet with 11-inch diameter pole pieces. Acceleration took place in the gap between a pair of hollow D-shaped electrodes with grids at the edges of the gap to confine the electric field to the space between the "dees." Lawrence had emphasized that it was important to have no electric field inside the dees.

A crucial step came quite by accident. During the summer, while Lawrence was away on a trip, Livingston decided to see what would happen if he removed the grids, since they necessarily intercepted some of the particles and tended to reduce the number of particles accelerated. Surprise! The resonance still worked and the beam intensity leapt up by a factor of 100 — far more than could be accounted for by the beam loss from the grids. The reason — quite unanticipated by Livingston and Lawrence, but recognized almost immediately — was that the electric field inside the dees produced electric focusing, which kept the particles from flying off to the top or bottom of the chamber as they had done before. This effect was most important at the center of the device.

An equally important development was the discovery of magnetic focusing. To smooth out possible imperfections in the uniform magnetic field Livingston inserted magnetic "shims," thin iron plates, between the magnet poles and the vacuum chamber. He found that he got the best beam intensity when these shims were shaped to make the field a little stronger at the center than toward the outside. After this empirical discovery it became clear that the curved lines of magnetic force associated with the field becoming weaker toward the outside would focus the beam, imparting a downward component of force to particles above the median plane and vice versa. And this effect becomes more pronounced toward the outside of the orbits, complementing the electric focusing, which is strongest at the center.

Stanley went to Cornell in 1934 where he was one of three authors on a series of papers that became the "bible" of nuclear physics. In 1938 he went to MIT and built a cyclotron that was heavily used during World War II for sterilizing blood and producing radioactive tracers for medical purposes.

During the war he worked on radar and radar countermeasures, in anti-submarine warfare. Soon after the war he took leave from MIT and was instrumental in the start-up of Brookhaven National Laboratory, where he was the driving force in building the first accelerator at that laboratory, namely the Cosmotron. Shortly after that, he suggested a modification of design that led to the concept of alternating gradient focusing. He then moved to Cambridge, Massachusetts where he constructed the 6 GeV electron accelerator, the Cambridge Electron Accelerator (CEA), and served as first director of the CEA for over 10 years. He retired to Santa Fe in 1970, where, amongst other things, he wrote a history of Los Alamos and, in his free time, produced jewelry.

Fig. 2.1 Here is a picture of the 11-inch cyclotron built by Lawrence and his graduate students, David Sloan and M. Stanley Livingston at the University of California, Berkeley during 1931. The machine gave out one billionth of an ampere of 1.22 MeV protons. It was the first cyclotron to exceed 1 MeV and at the time, Livingston recalled, "Lawrence literally danced around the room with glee."

chamber between the magnet poles which was low enough for particles not to be scattered as they were accelerated (see Fig. 2.1). Long afterwards Livingston wryly reported "Lawrence was my teacher when I built the first cyclotron — he got a Nobel Prize for it — I got a PhD."

II.2 Lawrence and the Early Cyclotrons

Encouraged by this first success, Lawrence set about building up a laboratory and a research team — perhaps the first on such a scale. To build accelerators, even these early cyclotrons, whose diameter is measured in inches rather than kilometers, required resources beyond those available in the normal university research departments of the 1930s. Lawrence started a process — unusual at the time but one with which modern physicists are only too familiar — of trawling for grants and support.

Lawrence wanted these funds for bigger and bigger cyclotrons. In January of 1932, using a larger (11-inch) cyclotron, Lawrence and Livingston, achieved 1.22 MeV.

Soon there was a laboratory and a 27-inch cyclotron, a 37-inch cyclotron, a 60-inch cyclotron (see Fig. 2.2), and then plans for a 184-inch cyclotron. Raising money, especially during the Depression and long before the federal government was financing scientific research, was a very difficult job, which was all-consuming for Lawrence. He was very skillful and persistent in pursuing possible sources of funds, including the physics department officials of the University of California. They were most supportive, but at some stage the pressure was clearly wearing, to the point that the chairman of the physics department, Robert Birge, said, "Berkeley has become less a university with a cyclotron than a cyclotron with a university attached."

Meanwhile cyclotrons were being built in many other places and Lawrence was very generous in sharing the knowledge developed in his laboratory, setting a fashion of collaboration which has benefited the accelerator world ever since. By 1940 there were 22 cyclotrons under development or completed in the US alone. This reinforced the support for his own research.

Fig. 2.2 The 60-inch cyclotron. The picture was taken in 1939 and the scientists about the machine are Cooksey, Corson, Ernest O. Lawrence (third from the left), Thornton, Backus, Salisbury, with Luis Alvarez and Edwin McMillan above.

All this was helped along too by Lawrence's own personality. Often discoveries are made by retiring and painstaking people and it is left to others to sing their praises but Lawrence was nothing if not an extrovert. He was ideally suited, not only to rouse support for bigger and better cyclotrons, laying the foundations of a Laboratory whose reputation is still the envy of the world; but also to inspire his coworkers to greater and greater achievement. Lawrence's leadership, characterized by ever-bolder ideas, dominated his Laboratory. To achieve this he developed ever-growing competitive research teams each working even harder than the other. Managing the skills of others became for him a major activity — presaging today's lab directors. This new way of doing science was, many think, Lawrence's primary contribution to the field. (See the sidebar on the Lawrence Berkeley National Laboratory as well as Fig. 2.3.)

A generation or more afterwards we come across those who were his students or taught by his students and who emulate his approach to science and to life.

Lawrence had a profound influence on the atmosphere at the Berkeley Radiation Laboratory. It had a spirit of dare-do, optimism and camaraderie that

has been rarely matched by any other laboratory. Los Alamos during World War II inherited it, but this was no accident as J. Robert Oppenheimer, who led Los Alamos, had been Lawrence's closest friend for the previous decade.

Lawrence's group enjoyed a commonality of purpose and sense of continued accomplishment that has rarely been duplicated elsewhere. Inevitably, there are many stories concerning the members of Lawrence's laboratory. Jackson Laslett married Lawrence's part-time secretary, Barbara Bridgeford. Glen Seaborg married Lawrence's full-time secretary, Helen Griggs. Ed McMillan married Lawrence's wife's sister.

Lawrence was a "hard driver." On two separate occasions he fired Robert Wilson, who was later to become one of the giants of accelerator physics. The first time this happened was when Bob, then a graduate student, while tinkering with the cyclotron, broke a seal, which prevented the cyclotron from working on the following day when Lawrence wanted to demonstrate it to a prospective donor of funds. He was re-hired thanks to the intervention of Luis Alvarez. The second time was a bit later — this time for melting a pair of pliers while probing inside the cyclotron. When offered his

Lawrence Berkeley National Laboratory

The Lawrence Berkeley National Laboratory was christened, in 1931, the Radiation Laboratory (the "Rad Lab") by its founder, Ernest Orlando Lawrence. His purpose in establishing the Lab was to pursue his invention of the cyclotron. This machine opened up a period of investigation in nuclear physics, biology, chemistry, and medicine that, arguably, has never been matched at any other laboratory. It led directly to seven Nobel Prizes in physics and chemistry. In later years the Rad Lab collected another three Nobel Prizes. These, though not directly related to the cyclotron and its descendents, might still be said to be the consequence of Lawrence's inspiration.

Lawrence originated the idea of scientific research carried out by individuals with different fields of expertise, working together as a team — a model now used in almost all fields of science and which is, perhaps, Lawrence's greatest legacy. The Rad Lab method was adopted for Los Alamos (where the atomic bomb was developed) — carried there by J. Robert Oppenheimer, who had been a leading member of the Rad Lab. It was also adopted at the MIT Rad Lab — a center of radar development during the war. After the war, Lawrence, with Edward Teller, founded the Lawrence Livermore National Laboratory — again, the Rad Lab method went with them.

Lawrence built a series of cyclotrons (described in some detail in the text) and then linear accelerators, synchrotrons (culminating in the Bevatron), and heavy ion accelerators (culminating in the Bevalac). Much science resulted from these machines, as is clear from all the Nobel prizes that users of these machines garnered.

As nuclear physics grew into particle physics and required accelerators too large for Berkeley, the Laboratory broadened out into other fields of science to become today's multi-purpose national laboratory. In 1971 it was formally separated from the Lawrence Livermore Lab. The Lawrence Berkeley Lab centerpiece is a very powerful x-ray source providing small, intense beams: the Advanced Light Source. Renamed the Lawrence Berkeley National Laboratory, it is part of the Department of Energy, operated, as has always been the case, by the University of California.

It leads in many fields of science and engineering research. The work is always unclassified, spanning a wide range of scientific disciplines with major efforts in fundamental particle physics, nuclear physics, and studies of the universe, quantitative biology, medicine, nano-science, new energy systems and environmental solutions, and the use of integrated computing as a tool for discovery. Out of this work have come many practical and immediate consequences as well as many contributions to basic science: for example, in medicine, pioneering the use of accelerators for treating cancer, the developments in PET scanning that were effectively transferred into clinical use, and understanding the difference between high density and low density lipoproteins. A second example is in conservation where the Laboratory has pioneered the analysis of, and proper design of, buildings, and developed high efficiency lighting.

The laboratory is perched on a 200-acre site in the hills above the University of California's Berkeley campus where it enjoys a commanding prospect of San Francisco Bay — an ideal place from which to view the universe. It has an annual budget of more than $500 million and among its staff of about 3,800 are more than 500 students.

---

job back, Wilson, who by now had his PhD, declined and went to Princeton. Nevertheless Bob Wilson had been profoundly influenced by Lawrence. Decades later McMillan's wife, describing their similar personalities, told one of the authors, "you might never have had the pleasure to meet Lawrence but you can be sure that Bob Wilson was Ernest Lawrence 'to a tee'."

On another occasion Lawrence saw someone with his feet up on the desk. He promptly went over and said, "You're fired," to which the victim replied, "You can't fire me; I work for the phone company." In spite of this rather summary approach to job appraisal there was tremendous loyalty amongst his staff and often genuine affection. Staff would meet every Monday evening at Lawrence's home. These evenings were devoted to a seminar and discussion of some recent exciting physics — not necessarily related to cyclotrons or nuclear physics. This tradition was kept alive by Luis Alvarez decades later and he was always proud to comment that it was in the tradition of Lawrence — only the venue had changed.

We have spent a long time describing the inventor of the cyclotron because his history is intimately bound

up with that of his invention and hopefully illustrates all that it takes to invent. Having said so many flattering things about the father of the cyclotron, and the laboratory he founded, it is only fair to remind the reader, now studying the anatomy of discovery, that all great men and great labs make mistakes. Even Homer nods! Lawrence was not anyone who might be put off by doubts or criticism and if at times this led him to make mistakes, he took these in his stride. In fact, Lawrence and his cyclotron laboratory missed out on two great things and were wrong about another. They were not put off by this and, in the long run, went on to bring greater glory to their laboratory than those that "beat them out."

The first setback was that Lawrence et al. were beaten in the race to produce nuclear reactions by J.D. Cockcroft and E.T.S. Walton in 1932, who were recognized for this achievement by the Nobel Prize in physics (1951) "for their pioneering work on the transmutation of atomic nuclei by artificially accelerated atomic particles." The second was something that with hindsight should not have been. In spite of Lawrence having the potential to produce induced radioactivity at Berkeley,

Fig. 2.3 An aerial view of the center of Lawrence Berkeley National Laboratory. It is located on a hill behind the University of California campus. One can see the landmark Campanile of the campus as well as the view in the distance of downtown Berkeley, and even the Golden Gate Bridge. In the center can be seen the circular building of the old 184-inch cyclotron — now housing an accelerator called the Advanced Light Source.

this was discovered by Frederic Joliet and Irene Joliet-Curie, who received the Nobel Prize in chemistry (1935) for their "synthesis of new radioactive elements." When he heard of this it only took Lawrence, and his people, half an hour to reproduce the Joliet-Curie result! They had been prevented from doing so by Lawrence's passion for saving money. He had hooked the Geiger counters to the cyclotron power switch so that any induced activity remained unrecorded when the cyclotron was off.

The other setback to his reputation came when Lawrence mistakenly proposed at a Solvay Conference in 1933, in which he was the only American, that the break-up of deuterium would be a power source. This statement brought upon him scathing comments from many of the great men of physics: Niels Bohr, Ernest Rutherford, Werner Heisenberg, and James Chadwick (all Nobel prizewinners) as well as other great physicists such as Otto Frisch. The reader may remember more recently that the proponents of "cold fusion" seemed to make the same mistake. Subsequent careful work disclosed that contaminated apparatus was the source of Lawrence's error.

It was left to Edwin McMillan, then a new member of Lawrence's Laboratory, and who would later make important contributions to accelerator science, to restore Berkeley's reputation. Ed was a very careful scientist. He extended the method used in the faulty work on deuterium to study heavier elements. He discovered the very important disintegration of deuterium (the nucleus of one proton and one neutron) — an effect that J. Robert Oppenheimer and his first student Melba Philips were the first to explain.

And so, despite serious setbacks, which would have discouraged lesser men, Lawrence went on, learned from his mistakes, and saw his Laboratory accomplish great things. Laboratory participants eventually collected ten more Nobel prizes. The Laboratory played a key role, either in developing or greatly extending, linear accelerators, synchrotrons, heavy-ion accelerators, nuclear medicine, bubble chambers, computational analysis in high-energy physics and many other things. The Radiation Lab served as the model for almost all high-energy physics laboratories, such as the Brookhaven National Laboratory, Stanford Linear Accelerator Laboratory, and Fermilab, in the US. Both CERN, founded in 1954 by 20 European countries; and DESY, Germany's national electron laboratory, used many aspects of the Berkeley model.

And now is the time to see how the principles of more recent accelerators came to be discovered in the course of the development of the cyclotron.

II.3 Transverse Focusing

Particles have a long and high-speed journey to make as they are accelerated and they may well first start out in a direction that takes them well away from their acceleration orbit and head towards the upper or lower poles of the cyclotron magnet. They therefore have to be persuaded to return to the radius at which they will be accelerated and away from the top and bottom poles — a process called transverse focusing.

Vital as this would appear to be, it was not even considered by Lawrence in his original proposal or in the first work with Stan Livingston. They were lucky that the gap between the poles was quite large and that in spite of losses, enough beam was accelerated to make their cyclotrons a success. The need to understand the focusing only came as they aimed to reduce the amount of beam lost. In fact, it was Stan who stumbled upon the

subject, by accident, but then quickly realized what was going on. This transverse focusing and its counterpart in the direction of acceleration have become extremely important as accelerators have become larger in radius and their magnets have become smaller in cross section to economize the cost of the ring. Many workers devote

their whole lives to aspects of focusing and we shall therefore return to it at greater length in other chapters.

There are two basic kinds of focusing depending on the fields: electrostatic and magnetic. Both were discovered, in the course of looking for other things, by Stan Livingston. In the earliest cyclotrons the open ends of the "Dees," which were like a cake tin cut across its diameter, were closed by a grid of wires. This was put in at the risk of scattering particles out of the beam, in the belief that it would produce a much more even accelerating field in the gap between the two halves of the tin and keep the beam in resonance. In the summer of 1931, while Lawrence was away, Livingston decided to see what would happen if he removed the grids. To his surprise, the resonance was still there and the intensity increased by a factor of 100. The reason, unanticipated but quickly appreciated, was that the curved field lines of the electric field spilled out from the open ends of the dees when the grids were removed. This provided vertical focusing and prevented the particles from flying off to the top and the bottom of the accelerating tank. This electrostatic focusing proved to be particularly effective at low energies near the center region of the cyclotron.

Magnetic focusing was also discovered in the fall of that year, as Livingston mounted magnetic shims, thin iron disc-shaped plates inserted between the main magnet and the vacuum chamber. His aim was to smooth out the magnetic field; but he discovered that he obtained the most intense beam when the field was not constant, but when these shims were shaped so as to give the strongest field in he center of the cyclotron and a slightly (only slightly) lower field at the outside. It was immediately clear that the bent magnetic field lines, like those of the electrostatic field were focusing the accelerated particles into the median plane. This magnetic focusing was most effective at high energy and provided the perfect complement to electrostatic focusing.

We will return to the subject of transverse focusing in Chapter IV. It is of vital importance for the design of these machines; Donald Kerst and Robert Serber, when building betatrons in 1941, first analyzed the effect quantitatively. A complete analysis of transverse focusing came in 1952 after another coincidental discovery, in which Livingston again played a major role.

II.4 Relativistic Limitation

By 1937 Lawrence, who by then had built a number of large cyclotrons of increasing size, set his sights on building one that would deliver 100 MeV protons. Hans Bethe and his student Morris Rose thought this was not possible. They wrote a paper in which they concluded that as protons began to feel the effect of relativity, the increase of their apparent mass would destroy the cyclotron resonance condition and make it impossible to go much beyond the 37-inch machine. The opinion of Bethe could not be lightly dismissed. He led the theoretical section under J. Robert Oppenheimer at Los Alamos and later won the Nobel Prize for his explanation of how the sun gave out solar energy in prodigious amounts. However, Lawrence responded to Bethe with the remark "We have learned from repeated experience that there are many ways of skinning a cat." The more circumspect Edwin McMillan did more careful theoretical work and convinced Bethe that protons might nevertheless be accelerated to at least 20 MeV and possibly 30 MeV.

Lawrence always argued that money, not relativity, was the primary problem. In November 1939 he was awarded a Nobel Prize and the fame greatly boosted his ability to raise money — and with it his ambition. Deuterons with twice the mass of the proton would be less prone to relativistic effects and would stand a chance of reaching his goal of 100 MeV. The cyclotron to achieve this would have a diameter of 184 inches (it would have been larger, but that was the largest size of commercially available steel plate); see Fig. 2.4. Just how it was to reach such energy was never made clear, but such energy would allow the artificial production of the new meson that had just been observed in cosmic rays. The cyclotron was soon started and commissioning took place in the summer of 1941.

About this time a solution to the relativistic energy limit did in fact emerge, thanks to a method of transverse focusing proposed by L.H. Thomas. This involved dividing the poles of a cyclotron into a number of segments (like an orange) and shaping their boundaries to restore particles which swung away from a circular orbit. The theory was difficult to understand and it was only much later, after World War II, and after some other less elegant solutions had been tried, that it was applied. Thomas focusing was studied at Berkeley by Oppenheimer's student, Leonard Schiff, who later made many important contributions to relativity, did some extensive calculations concluding that implementation of Thomas' ideas was not easy. It was just too much ahead of its time.

II.5 Calutrons

By 1940 scientists were becoming more and more

Fig. 2.4 The magnet of the 184-inch cyclotron. This magnet was first used to develop the calutron. After the war the 184 magnet was incorporated into the largest cyclotron that Lawrence ever constructed and became the source of much very important physics.

Llewellyn H. Thomas (1903–1992)

British physicist who emigrated to the US in 1929
Inventor of the Thomas cyclotron

Born in London, he was educated at Cambridge University under Ralph Fowler. As a graduate student he won a one-year fellowship to the Neils Bohr Institute in Copenhagen. During that year — while still a graduate student — he produced two monumental works in physics. One was a simple statistical model of heavy atoms. The mathematics of this model, the Thomas-Fermi Model is simple, and the model is still used even to this day. The other work was a realization and analysis of the consequence of special relativity on a spinning electron. Here the analysis is very difficult, but the result is a simple factor, the Thomas factor, that modifies the magnetic moment of an electron. He was elected a Fellow of Trinity College (Cambridge) while still a graduate student.

In 1929 he moved to Ohio State University, and in 1938 he invented the Thomas cyclotron. In 1943 he moved to the Ballistics Research Laboratory at the Aberdeen Proving Ground where he worked, during World War II, on matters of ordnance, such as shell fragmentation and lateral motions of spinning shells. After one year back at Ohio State, in 1946 he moved to the new IBM Laboratory, at first in, and then near, Columbia University (New York) and in 1950 he became officially associated with Columbia. At IBM he used his interest in technology and vast mathematical powers to make many important contributions in the early days of computers. He was also widely consulted by the Columbia physics faculty and was known as "the sage of 116th Street." In 1968 he retired from IBM and Columbia University, but took a post-retirement position at North Carolina State University, where he remained until 1976.

He was reserved and tended to work by himself so he had few collaborators and few students; although a very notable one — Leonard Schiff — went on to make important contributions to general relativity. He was, however, most approachable and generous with his time and had a fine sense of humor.

involved with the war and weapons research was starting in a significant way. At first Lawrence was reluctant to become involved, but members of his laboratory were one by one joining in the war work and even leaving to join dedicated laboratories. Ed McMillan and Luis Alvarez left for the MIT Rad Lab to work on radar in the fall of 1940 but Lawrence had started construction of the 184-inch in the summer of 1940 and was reluctant to see it interrupted.

However, in the spring of 1940, Lawrence became convinced that an atomic bomb could be, and should be, built. In the context of the time the fear that Hitler might get there first was extremely persuasive. By 1941, Lawrence became quite involved with the US planning, and more and more frustrated with the slowness of the US effort. He quickly focused upon the crucial point of separating uranium. Natural uranium, consisting mainly of the isotope U238, is unsuitable for a bomb since it contains less than 1% of the fissionable isotope, U235. Increasing the percentage of U235 by separating it from other isotopes was the crucial point.

Separation cannot be done chemically, but it can be accomplished by at least three other methods, all based on the half of one percent difference in mass of the two nuclei. The first, called "electromagnetic," relies upon a magnetic spectrometer that bends the different isotopes by different amounts (the difference is very small indeed). The second is to exploit the different rates of gaseous diffusion of (say) uranium hexafluoride. A third method uses centrifuges. A fourth method, which only became available long after World War II, is to use a laser that excites atoms of one of the isotopes, and not the other, because of the slight difference of atomic energy levels.

Alfred Nier had started to use a mass spectrograph to do the electromagnetic separation, but the rate was much too slow (by factors of a million). Lawrence intended to use the cyclotron (really only its magnet) and vastly speed up the process. By December 1941 his people had already converted the 37-inch to separation and produced a beam of uranium ten times as powerful as Nier's. Although, Lawrence made many unsupported statements and was — by any reasonable standards — unduly optimistic, in the long run he was successful. In 1942 he converted the magnet of the 184-inch cyclotron (by then only partially completed) into an electromagnetic separator. It was named a "calutron" in honor of the University of California.

The actual production of separated uranium was done at Oak Ridge at the Y-12 plant, which had 500 calutrons based on Lawrence's machine (see Fig. 2.5).

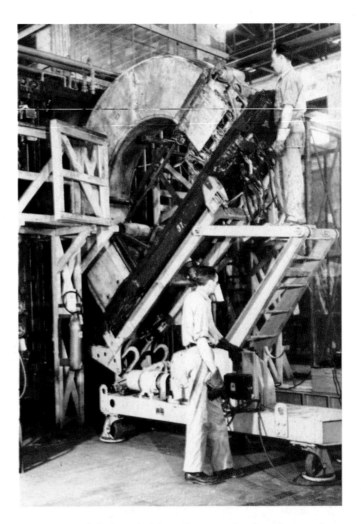

Fig. 2.5 The "C" shaped alpha calutron tank. The calutrons (here an early version is shown) were used for separating the isotopes of uranium to supply the nuclear material for the atomic bomb dropped on Hiroshima.

With an unbelievable amount of effort by Lawrence and his men, the plant produced significant amounts of separated uranium and, eventually, all of the material for the Hiroshima bomb simply because other separation methods were not brought on-line in time. Since World War II the other methods have proved cheaper and more convenient than the electromagnetic method.

II.6 Cyclotrons for Peace Again

The World War had affected Lawrence's laboratory considerably. His largest, 184-inch, cyclotron had been diverted from its original purpose until after the war and cyclotron development had ground to a halt as Lawrence's team was dispersed. However preoccupied they may be with other more pressing business, there is usually a small region in the back of a scientist's mind that always works away subconsciously — on the lookout for another "Eureka."

In 1945, Ed McMillan, who was still at Los Alamos, suddenly realized an ingenious way to allow cyclotrons to overcome the relativistic limitation and accelerate to much higher energies. His idea was to interrupt the steady stream of particles accelerated in the cyclotron, and send pulses, or bunches, of particles on their journey from the centre of the machine to the rim of the magnet. By tuning the frequency of the radio energy source one might hope to keep in resonance as the revolution frequency of the particles sagged due to their relativistic increase in mass.

The final piece of the jigsaw was to keep the particles in synchronism with the RF by "phase focusing." By accelerating the ideal particle on the falling edge of the RF sine wave, particles which lag behind get more energy and catch up with their brothers while those that are too far ahead see a smaller accelerating voltage. The same idea was hit upon, independently by V. Veksler.

In fact phase focusing is at the heart of the synchrotron (see Chapter V), a machine which was invented about this time and, once demonstrated on a "synchrotron model," was soon to be incorporated into the 184-inch cyclotron.

II.7 FFAG

We will read in Chapter V that the discovery of the synchrotron in the closing days of the war dominated the construction of new proton machines in the decade that followed — but it was not to the complete exclusion of cyclotrons. With the discovery in the early 50s by Courant, Livingston and Snyder of alternate gradient focusing (described in Chapter V) all sorts of new focusing schemes became possible. This was just at that time that the Midwestern Universities Research Association (MURA) was formed to urge the Atomic Energy Commission (AEC) to build an accelerator in the Midwest. Up until that time only the west and east coasts boasted accelerators. The MURA Group was led by Donald Kerst of betatron fame, and made many innovative contributions to the physics of beams, but sadly they were never able to obtain AEC support for construction of a large accelerator. In due course, MURA reemerged as the Physical Sciences Laboratory at Wisconsin and became one of the early players in developing synchrotron radiation science. However, in a certain sense the goal of MURA was achieved, for an accelerator, the Zero-Gradient Synchrotron (ZGS), was built at the Argonne National Laboratory.

Keith Symon, then a member of MURA, realized that an accelerator could be constructed with a mag-

Vladimir Iosifovich Veksler (1907–1966)

Soviet physicist
Inventor of Phase Focusing
Lenin Prize, 1959
Atoms for Peace Prize, 1963
Member of the USSR Academy of Sciences, 1958

He was born in Zhitomir and educated at Moscow State University, receiving a Diploma in Electrical Engineering in 1931, a Candidate Degree in Physico-Mathematical Science in 1934 and Doctor of Physico-Mathematical Science in 1940.

He worked at the All-Union Institute of Electrical Engineering from 1930 until 1936, the Lebedev Institute of Physics from 1936 to 1956, and then, for the rest of his career, at the Joint Institute for Nuclear Research in Dubna. This Institute was jointly supported by all of the nations in the Soviet bloc and was in some ways a smaller version of CERN. At Dubna he became Director, but always maintained his ties to the Lebedev Institute and taught at Moscow State University. He studied cosmic rays while at the Lebedev and joined expeditions to Central Asia. His most important contribution was the independent discovery of phase focusing in 1944. This discovery led to a number of accelerators in the Soviet Union including the 10 GeV Synchrophasotron at Dubna, constructed from 1953 to 1957. As director of Dubna he developed the first laboratory for high-energy physics in the Soviet Union and made sure that this laboratory had an appropriate infrastructure and both accelerators and detectors.

He was a quiet, modest, person with a good sense of humor. He was a strong advocate of international amity amongst scientists (at a time when this was quite a radical concept).

netic field which, like that of a cyclotron, and in contrast with that of a synchrotron, was constant in time. The field might be shaped to accommodate a wide range of energies from injection to top energy within a relatively small band of radii — its magnet aperture rather like the brim of a sombrero. The beam would be accelerated in a series of bursts relying upon McMillan's phase focusing but it would not be necessary to pulse the magnet. These accelerators were called "fixed-field alternate gradient" (FFAG for short); see Fig. 2.6. There were a number of ways to accomplish this, all of which differed according to the pattern of the magnetic field around the circumference, and MURA constructed models of both a "radial sector" and a "spiral sector" type. The spiral sector was more complicated, but more efficient in its use of magnetic field; see Fig. 2.7.

The reader will remember that Llewellyn Thomas had earlier also developed an azimuthally varying

MURA: A Flash in Accelerator History

The idea of an association called MURA sprang up from a general concern among high energy physicists in the Midwest, that accelerators were being built on both coasts (the Bevatron in Berkeley, the Cosmotron and the AGS in Brookhaven) but nothing was being built in the center of the U.S. Ten Mid-Western universities combined forces in September of 1954 to formally incorporate Midwestern Universities Research Association. Donald Kerst was appointed technical director, charged to design, obtain funding for, and build, a Midwestern accelerator. He had in fact been acting in this capacity since the summer of 1953 and surrounded himself with a number of young people, almost all of whom had no prior experience in accelerator design, but had enthusiasm and imagination.

In the summer of 1954, Brookhaven's discovery of alternating gradient focusing had triggered Keith Symon and Donald Kerst at MURA to invent a new kind of accelerator. The young Keith Symon had no prior accelerator experience, but knew a great deal about dynamics. They called their new concept a fixed-field alternating gradient (FFAG) accelerator. In such a machine the magnetic field does not vary in time and particles are accelerated and spiral outwards so that the full energy range from the injection energy to the final energy can be kept within the poles of a single magnet. Subsequently, Kerst invented a more efficient form of FFAG. This involved spiral magnets — a concept that has since been incorporated in many "spiral sector cyclotrons." We describe in Section II.7 more details about the FFAG idea and point out that the FFAG concept, but not the spiral concept, was independently discovered by Ohkawa and Kolomenski.

In late 1954 at Michigan, MURA started to build a small electron model of the simplest version of an FFAG: the so-called radial sector FFAG. Subsequently, a second model, this time of a spiral sector configuration was constructed, mostly in Illinois. Both the "Michigan Model" and the "Illinois Model" were moved to Madison, Wisconsin where MURA established its permanent location.

In 1955, MURA proposed the construction of a 20 GeV FFAG to the AEC. It would cost $17 million and take 5 years to construct. For reasons which we explain later, the AEC had other priorities and approval was not immediately forthcoming.

Meanwhile the FFAG concept gave Donald Kerst the idea that it might be possible to accelerate and store a sufficiently intense beam to make a collider. The MURA group developed the idea of piling up several beams side by side — a process we now call "stacking." Studies were carried out to demonstrate the reality of stacking both on the Illinois model and on a third model: their "Two-Way Model." Thus MURA was the first to develop a practical method of achieving colliding beams. Colliding beams were destined to become the basis of essentially every modern high-energy machine.

An accelerator with colliding beams was proposed (the "green book") in 1958 to the AEC. This was followed with a proposal (the "blue book") for an intense single beam machine, and again by another proposal for a higher energy single beam machine (the "gold book"). However by this time ordinary synchrotrons were developing rapidly and soon able to match the intensity of FFAG machines and feed storage rings which allowed colliding beams without the use of FFAG. Sadly, but inevitably, the MURA's FFAG proposals were rejected — and for good technical reasons.

With hindsight it is clear that the MURA group had pitted itself against an overriding interest on the part of the AEC not to start another laboratory in the Mid-West, but to strengthen the Argonne Laboratory; MURA for its part refused to cooperate with Argonne. At the same time, in the context of the cold war, the AEC urgently wanted a machine of higher energy than the 10 GeV Synchro Phasotron at Dubna and could not wait until 1960, or later, for the AGS at Brookhaven to be completed. (In fact the AGS was completed before the ZGS.)

The outcome was that the ZGS, at Argonne, was authorized by the AEC in 1957, and this sounded the death knell of the MURA in high-energy physics. Donald Kerst left that very same year, and MURA was officially dissolved in the early 1960s. Subsequently, the University of Wisconsin took over the MURA facility in Stoughton and used it as a most successful source of synchrotron radiation.

Besides FFAG, spiral sector cyclotrons, colliding beams, and stacking, the MURA Group made many important technical contributions to accelerator physics — contributions that have underpinned the ongoing development of accelerators. The Group was probably the most inventive in the history of accelerators, making contributions scarcely paralleled in accelerator history in their number and significance. But, despite MURA's inventiveness, they were never able to produce a proposal that was competitive with other concepts.

magnetic field that allowed cyclotrons to go beyond the relativistic limit even when run for a continuous beam. His work came long before the concept of alternating gradient focusing, but the importance of his work was unappreciated and even unknown to essentially everyone involved in accelerator design. His "Eureka" had fallen on deaf ears, which was a pity because it had offered a path back to the continuous beam machine. The MURA Group was to follow the same path.

At about the same time as Symon's innovative work, a similar idea was put forward by a young Japanese, Tihiro Ohkawa. The MURA Group learned of Ohkawa's work and invited him to join the effort, which he did (see sidebars for Thomas, Symon, and Ohkawa). Independently, A.A. Kolomenski, in the Soviet Union, also invented the idea of FFAG.

The MURA Group became somewhat obsessed with the FFAG concept and proposed a number of large accelerators in the range of 10 GeV based on this principle. Their virtue was their high intensity. None of these devices were ever realized.

II.8 Spiral Sector Cyclotrons

The work on FFAG, at MURA, aimed to provide an

Fig. 2.6 A visit to the MURA laboratory in Madison by Nobel prizewinner Niels Bohr in 1958. Overlooking the Radial Sector Model are (left to right): Robert Haxby, Ragnar Rollefson, Harrison Randall, Subrahmanyan Chandrasekhar (a future Nobel prizewinner), Niels Bohr, Charles Pruett, and Lawrence Jones. The model was built by Lawrence Jones and Kent Terwilliger, of the MURA Group, in Michigan in 1955. There are 8 sectors, and electrons of 30 keV were injected at a radius of 34 cm and accelerated by betatron action, to 400 keV at a radius of 50 cm.

Fig. 2.7 The Mark 2 version of a spiral sector FFAG built by the MURA Group in Wisconsin from 1956 to 1959. It had 6 sectors and accelerated electrons by betatron action from 35 keV at an injection radius of 31 cm, to 180 keV, at 52 cm radius.

Keith R. Symon (1920–)

American physicist
Inventor of Fixed Field Alternate Gradient Accelerators (FFAG)
American Physical Society R.R. Wilson Prize, 2005

He obtained a PhD in 1948 from Harvard University. After that he was on the faculty in the Physics Department of Wayne University, Detroit, from 1947–1955. Then he joined the faculty of the Physics Department of the University of Wisconsin-Madison. In 1990 he was made Emeritus Professor of Physics.

At the same time he was a staff member of the Midwestern Universities Research Association (MURA), 1956–67 and then technical director, 1957–60. He was chairman of the Argonne Accelerator Users Group, 1961–62; acting director, Madison Academic Computing Center, 1982–83; and acting director, UW-Madison Synchrotron Radiation Center, 1983–85.

He has been a most productive research physicist working in the areas of the design of particle accelerators and plasma physics. Besides inventing FFAG accelerators he developed the smooth approximation method for approximating the solutions of differential equations with periodically varying coefficients (called the "smooth approximation"), formalized the theory of radio-frequency acceleration in fixed field accelerators, and contributed greatly to the development of colliding beam techniques. He was the first to develop the theory of collective instabilities in accelerators and plasmas (a subject that spawned a thousand papers). He also contributed to the linearized analysis of inhomogeneous plasma equilibria and developed a method of bit pushing and distribution pushing techniques for the numerical solution of the equations employed in both plasma physics and the study of collective instabilities in accelerators (the Vlasov equation).

He was an outstanding supervisor of graduate students, having been the major professor for 20 graduate students gaining PhD's from the UW-Madison. He was author of *Mechanics*, a popular undergraduate textbook (Addison-Wesley, 1953, 3rd ed., 1971). He married in 1943 and he and his wife had 4 children. They now have 7 grandchildren. He was one of the organizers, president and treasurer of the Spring Green Literary Festival. He was an avid skier, downhill and cross-country, and engaged in both wilderness and white water canoeing.

Tihiro Ohkawa (1928–)

Japanese physicist, who worked in the United States after 1955
Inventor of Fixed Field Alternate Gradient Accelerators (FFAG)
Recipient of the Maxwell Prize from the American Physical Society

His first exposure to physics was when he was recruited, at age 16 during World War II, to help in Prof. Yoshio Nishina's cosmic ray laboratory. He received his Bachelor's degree and his Doctor of Science Degree from the University of Tokyo. Subsequently he joined the MURA Group in the mid-west of the United States. He spent some time at CERN in Europe, but most of his years were spent doing plasma physics at General Atomics, during which time he was also a professor at the University of California in San Diego. At present he is at Archimedes Technology, which he co-founded.

He has worked, most productively, in the fields of nuclear physics, fusion energy and plasma physics, plasma processing and biotechnology. Besides his major contribution to accelerators he has, in fusion and plasma physics, contributed experimentally to the important problem of plasma transport; i.e., the diffusing of hot plasma out of a fusion device. He developed non-circular cross section tokamaks which greatly increased the plasma pressure that could be confined by an external magnetic field. He has also worked on the non-inductive methods of generating current in toroidal plasma (so as to allow nuclear reactions). His work has led to over a hundred patents in plasma devices and biotechnology. He initiated the US/Japan fusion cooperation program and co-founded the Archimedes technology for nuclear waste disposal.

He is very interested in all sorts of inventions, has even been characterized as an "invention junkie." His hobbies include skiing, tennis and golf, and generally he is known for his debonair ways and is reputed to be a "man about town."

---

alternative to the synchrotron for high-energy physics and they were thought to be particularly appropriate for colliding beam devices. However, it became clear, even at an early stage, that the spiral sector concept would also make a very good cyclotron. In particular, the varying field allowed separated sectors which could be relatively easily manufactured and assembled. Most importantly, in contrast to the pulsed phase focused cyclotrons, the beam remained much closer to resonance and the radio frequency changes were much smaller. Soon, a number of spiral sector cyclotrons were under study and construction, notably at Oak Ridge (USA), Chalk River (Canada) and Rutherford Laboratory (England). Later the first superconducting cyclotron was built at Michigan State by Henry Blosser, who was associated with MURA while he was a student.

Fig. 2.8 Canada's Tri-University Meson Facility, TRIUMF, the world's largest cyclotron. In this illustration a worker is checking the accelerator for worn or damaged components. The diameter of the machine is about 18 m and ions travel a total of 45 km and, as they spiral outward, are accelerated to 520 MeV. The machine started in 1974 and is still in operation.

II.9 Modern Cyclotrons

Ordinary cyclotrons are used mainly for dating purposes (as discussed below), and isotope production. However spiral sector cyclotrons are important, even to this day, in nuclear physics. Three very large and very productive machines are the 88-inch cyclotron at Lawrence's Berkeley Laboratory, the 520 MeV cyclotron at TRIUMF (Fig. 2.8), and the 590 MeV cyclotron at the Paul Scherrer Institute, Switzerland; these are also major cyclotron labs at IUCF and RIKEN.

In recent years a few 200 MeV proton machines have also been built for cancer therapy.

II.10 Applications

A major application is in medicine as described in Chapter IX. Another kind of cyclotron is used these days for radioactive carbon dating. Radioactive dating of once living organisms works because as time passes cosmic rays turn some of the atmospheric nitrogen into radioactive carbon, C14, by capture of slow neutrons which decays slowly, with a half life of 5,370 years.

(After 5,370 years half of the radioactive carbon has disappeared.) Once an organism dies, it no longer takes up carbon dioxide, and thus by measuring how much radioactive carbon is left in the organism, one can tell how much time has passed since it died.

Radioactive dating has been a very important matter for archeologists, anthropologists, and environmentalists. However, rather large samples were needed for the analysis and in some cases, such as dating the Shroud of Turin, taking a large sample was ruled out. However in the 1970s Luis Alvarez (characterized as "the greatest experimental physicist that ever worked in Lawrence's laboratory") and the much younger Richard Miller suggested, and then developed, cyclotron radioactive dating. This uses only a minute sample. The idea was to accelerate a small amount of carbon and, then, use the slight difference of mass between radioactive carbon and ordinary carbon to separate them by tuning the cyclotron to each in turn. Thus the total number of radioactive atoms are counted; not just the ones decaying. The increase in sensitivity is many thousand fold.

Fig. 2.9 A modern very small cyclotron that has many different uses (such as dating) which are described in the text. The cyclotron has a permanent magnet designed by Klaus Halbach (center) surrounded by the team that constructed the machine.

Subsequently, in 1994, a group at the Lawrence Berkeley laboratory built a small cyclotron (the size of a microwave oven) for this purpose. The cyclotron had a number of innovations including the use of permanent (samarium cobalt) magnets (rather than the electromagnetic magnets, with their large power supplies). Its small size and portability has made it extremely useful. (See Fig. 2.9.)

Chapter III. Linear Accelerators

III.1 Science Motivation — An Idea in Search of a Technology

The invention of the cyclotron led immediately to a series of highly successful accelerators with energies far beyond the reach of electrostatic machines. This was only possible because the technology needed to build the magnet and radio frequency systems for cyclotrons was ready and waiting to be used.

In contrast, the linear accelerator (or linac), a concept that actually predates the cyclotron by six or seven years, had to wait until 1937 before the klystron — the high radio frequency technology that it needed — was proposed. (See sidebar for klystrons later in this chapter.) It then took until 1946 before Alvarez invented an efficient design for a proton linear accelerator. In his design, acceleration was applied by a voltage wave traveling along the length of a waveguide, which was modified to slow down the wave to keep it in step with the particles. At about the same time, Hansen developed the electron linac. Only in the late 1940s and 1950s were linacs put to work for medicine and particle physics.

The linac idea had actually emerged 30 years earlier, in 1924, when it was proposed by a Swedish physicist, Gustav Ising. It was then the first practical idea for acceleration by the accumulation of a series of small steps of modest voltage. This seemed to avoid the breakdown problems of electrostatic machines and opened the door to all the accelerators that have been built in the eighty years that followed.

In Fig. 3.1, from Ising's original paper, we see electrons passing from left to right through a straight vacuum tube made of some non-conducting material such as glass. Traveling down the tube, they thread their way through three cylindrical metal "drift tubes." As they pass from one tube to the next, their energy is increased by a rather modest voltage, V, and this energy accumulates as they pass through more gaps. Only two gaps are shown but the idea can be extended so that

if there are n gaps the particle reaches an energy of n times V but — and here is the crucial point — there is no point in the apparatus which is more than this modest voltage V above ground potential.

Ising also shows the circuit for the charging system. A capacitor is charged up until it breaks down across a spark gap. The sudden change in voltage sends a pulse along cables to the drift tubes, arriving at each tube in turn. The timing is crucial, and the machine will only accelerate if the pulse is applied while the particle is shielded from any accelerating or decelerating field by the drift tube, which acts as a "Faraday cage." As the particle emerges from each tube, it should experience the full field between the tube it has just left, fully charged to voltage V, and the one it is approaching, which is still at zero potential. The polarity is chosen of course so that the field accelerates the particle: an electron approaching a positive tube will be accelerated to the next, whereas a negative tube would have the same effect on a positive particle. In the diagram we see the connections to each tube have different lengths to ensure this synchronization of beam and pulse.

Once the particles reach the next tube the process is repeated *ad infinitum* — or would be, were it not for the fact that the tubes have to increase in length as the particles travel faster. Only when they are close to the velocity of light will the particles go no faster and the tubes will get no longer, but by then each tube may be a kilometer long!

In 1928 Ising's idea was taken up by Wideroe who, frustrated by his own inability to make a practical version of the betatron, first demonstrated how it might work. (See sidebar for Wideroe.) Instead of Ising's pulses, Wideroe used oscillating voltages from a radio frequency oscillator to apply synchronized voltages to the drift tubes. Wideroe's model had only one powered drift tube but used the gaps between this and two grounded "dummy" drift tubes to accelerate at both

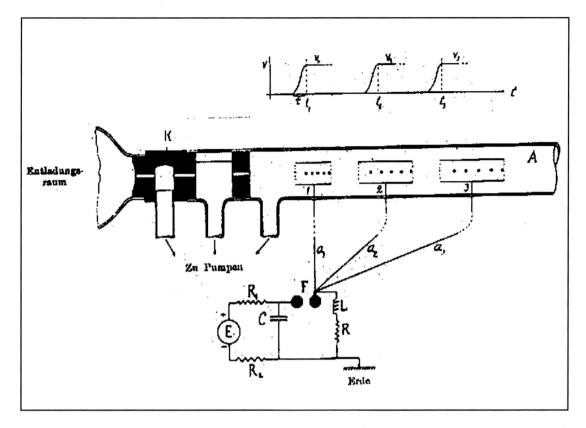

Fig. 3.1 A drawing, from Ising's original paper of 1924, showing his idea for an RF accelerator. Later Wideroe was able to turn this idea into reality, demonstrating RF acceleration for the first time and opening the door to all modern accelerators.

the positive and negative swings of the voltage wave. Wideroe's work set others off in the direction of constructing multi-drift-tube linacs and of course his work sparked off the idea of the cyclotron in Lawrence's mind (see Chapter II).

In those days radio had just appeared for the first time in people's homes and transmissions were at the low frequencies that corresponded to wavelengths of several hundred meters. Linac builders had to use the same low frequency radio transmitters as commercial radio. The distance between the accelerating gaps in Wideroe's type of linac must match the distance traveled by the particle in half a swing of the radio frequency.

Wideroe accelerated low energy particles (sodium ions of a few keV), which would only travel a few centimeters in the time it took his radio oscillator of 100 kHz to change polarity. However, when the aim is to accelerate above a few MeV and closer to the speed of light, the total distance particles must travel becomes very large compared with laboratory space available. Close to the speed of light they would have to travel hundreds of meters from one drift tube to the next — half the wavelength of radio transmitters in those early days of broadcasting. Thus for many years the cyclotron proved to be the more popular alternative to the linac.

Much later a tube called the klystron, invented in 1937 and used for radar in World War II, made power at higher frequencies more accessible. High frequency power tubes — first for 200 MHz and later in the gigahertz (GHz) range — became available, and linacs for relativistic particles became feasible. At a frequency of 1 GHz the wavelength shrinks to 30 centimeters, scaling down the length of the linac by many orders of magnitude. In the 1950s, this technology was incorporated into linacs, and the betatrons of the 1940s (see Chapter IV) were superseded by much larger and higher energy electron linacs.

That Ising's brilliant idea had to wait more than two decades is sad. It is an excellent example of an inspired invention that had to wait for a new technology before it could be put into practice.

Until the 1950s, linear accelerators were used almost exclusively for cancer therapy and nuclear physics research. There was one exception — Lawrence's Materials Test Accelerator (MTA) for manufacturing uranium — a huge machine that we shall describe later. More recently, as will be described in more detail in Chapter X, many other applications of linacs have emerged, such as neutron sources, drivers for free electron lasers, implantation accelerators for introducing specified ions into semiconductors, accelerators for "burning up" nuclear waste and accelerators for driving power reactors. It is rather ironic that nearly a century after Ising we are looking forward to using his invention as the only practical way to make a high energy linear collider — again we have to wait for the technology, either superconducting cavities or the two-beam concept of CERN's Compact Linear Collider (CLIC), to catch up with this simple concept — but we are almost there!

III.2 The Early Linear Accelerators at Berkeley

Linac development had not been completely dormant in the cyclotron era. In 1930 David Sloan and Ernest Lawrence (of cyclotron fame) built a linear accelerator for ions, using the ideas developed by Wideroe. Their initial linac reached 90 keV: eight times the voltage on any one drift tube. Then 13 more electrodes were added to reach 200 keV with only 10 kV on any one electrode. They had set out to "study the properties of high-speed ions," but never actually did that. Instead Sloan built an even larger linac, with 30 drift tubes, and accelerated mercury ions to 1.26 MeV. These machines had a wire mesh or grid structure at the downstream end of each drift tube to modify the field lines between the tubes and provide transverse focusing. Of course these grids obstructed the accelerated beam and produced beam losses, but without them the natural fields between drift tubes would have defocused the beam and produced even bigger beam losses.

Sloan then became deeply involved with the generation of x-rays by means of a resonant transformer device. This is not to be confused with Wideroe's beam transformer, which we discuss in Chapter IV. In essence it is a normal high voltage transformer but with the secondary winding ending in an open circuit gap in an evacuated tube. Electrons are drawn from one side of the gap by the strong electric field and accelerated to the other side of the gap, where they produce x-rays as if it were the anode of a normal x-ray tube. Sloan was able to reach 800 keV in the laboratory, and wanted to continue to improve the device so as to get to 1 MeV, but he was discouraged in this desire by the oncologists' advice that 500 keV was quite adequate for treating patients. Eventually these devices were built commercially and installed in a number of hospitals. They brought in a small amount of money to Lawrence's lab, enough to cover half the salary of both Stan Livingston (until he left in 1934) and David Sloan (until he left in 1937). After this there was no further linear accelerator work,

Rolf Wideroe (1902–1996)

Norwegian electrical engineer and physicist
American Physical Society R.R. Wilson Prize, 1992
Originator of many ideas and builder of the first practical linac

Norwegians feature prominently among those who have made major contributions to the progress of accelerators, but the intellectual contribution of Rolf Wideroe is quite outstanding. His interest in finding a better means to accelerate was aroused when, at seventeen, and still a high school student in Oslo, he read of Rutherford's groundbreaking discoveries of nuclear structure using only alpha particles from radioactive decay. His illustrious career started at Karlsruhe Technical University in 1923 lasting until his retirement in 1969 from ETH Zurich.

In the ten years from 1933 he moved out of accelerators and worked for Siemens and was concerned with the design of relays for high voltage equipment. His work there was sufficiently significant for the Allies, at the end of World War II, to imagine that he was contributing to Hitler's V-rocket program. He was imprisoned for several months in Norway before Odd Dahl secured his release.

During the war years he had indeed been working in Berlin on betatrons, which he was told would serve as a compact x-ray source in field hospitals. (Actually, the Nazi leadership hoped to use it to blind bomber pilots.) He agreed to do this when promised the release of his brother, Viggo, who had been imprisoned by the Nazis on suspicion of spying in Norway.

He is variously credited with the invention of the betatron, the linac, the synchrotron, and storage rings for colliding beams, and certainly he built the first pair of linac drift tubes for his Dr. Ing. thesis at Aachen in 1927 — a publication that Lawrence said gave him the idea for the cyclotron.

His many other claims as an innovator, though substantiated by patents, were often hidden from the free world of science by the clouds of war and had to be re-invented either independently or subsequently by others. It does seem clear, however, that his enthusiasm for storage ring colliders prompted his good friend Touschek, himself imprisoned by the Nazis, to build the first such device after the war.

for it was overshadowed by cyclotron activity. Nothing much more was done on linacs until the discovery of the klystron was followed by the arrival of high frequency technology after World War II.

III.3 Proton Linacs

A decade later, and just after World War II, Luis Alvarez invented a new linear accelerator that was very similar to that of Sloan and Lawrence, but could use the powerful high frequency radio sources developed for radar. (See sidebar for Alvarez.) There was a novelty in

Alvarez' structure. There were drift tubes, as in Ising's drawing, but they were suspended, not in a non-conducting tube, but in a copper cylinder. Together with the drift tubes, this formed a resonant electrical cavity in which the RF waves propagated — a kind of waveguide. In other early linacs, there were grids to provide radial focusing placed at both ends of the drift tubes. This device, now known as a drift tube linac (DTL), was designed originally to use the power provided by a 200 MHz radar unit, but in the long run the radar unit was not used. The first proton linac constructed by Luis Alvarez, a 32-MeV drift tube linac, became operational in 1948. (See Fig. 3.2.) A year earlier, construction had started near the town of Livermore on what would be the largest (and most short-lived) linac ever undertaken.

(See Fig. 3.3 and sidebar for MTA.) This project was named the Materials Testing Accelerator — a deliberately deceptive name. The MTA, built by E.O. Lawrence and Alvarez, was intended to be a device which would work in conjunction with a reactor (never built) to breed plutonium and tritium. It was a time when the US still appeared to have an inadequate supply of uranium ore either for nuclear power plants or for nuclear weapons. The initial component of this monstrosity was a linac with a vacuum tank 18 m in diameter and 400 m long; an airplane is reputed to have flown through it! It was designed to accelerate a third of an ampere of deuterons to 500 MeV. The finished MTA was expected to yield about 2 kg of neutrons per year, producing fissile fuels at a cost comparable to that from natural uranium.

Luis W. Alvarez (1911–1988)

American physicist
Recipient of the Collier Trophy
National Medal of Science
Enrico Fermi Award of the US Government
Michelson Award
Member of the National Academy of Sciences
Nobel Prize, 1970

Luis Alvarez made contributions in many different fields of physics, from his earliest days until he was well past retirement. He was, surely, one of the great experimental physicists of the last century.

He was born in San Francisco and started to study chemistry as an undergraduate at the University of Chicago, but quickly shifted to physics after taking a course in experimental physics, where, as he later put it, "It was love at first sight." His thesis work was on cosmic rays and, in particular, the difference in flux from the East and West. In 1934 he joined the Radiation Laboratory at the University of California in Berkeley and, except for the war years, remained there as a Professor of Physics all his life.

In the years before World War II, Alvarez was particularly prolific, having discovered the capture of electrons in beta decay (K-capture), determined the stability of He^3, and measured, with Felix Bloch, the magnetic moment of the neutron.

With the start of World War II, Alvarez went to the Radiation Laboratory at MIT, where he worked on ground-based radars. He invented the VIXEN method for detecting enemy submarines and, perhaps most importantly, Ground Controlled Approach radar, which is the basis for all such systems in use to this day for the safe landing of airplanes throughout the world. Shortly after, Alvarez went to Los Alamos, where he developed the shock wave method of measuring the strength of nuclear explosions. He was, in fact, on the Hiroshima mission, in order to measure the strength of the bomb.

After the war, Alvarez returned to Berkeley where he built the first modern, and practical, proton linear accelerator. During the Korean War, fearful that the US uranium supplies would be cut off and believing domestic supplies were inadequate, E.O. Lawrence conceived of a monstrous accelerator in order to breed plutonium (rather than in a reactor). Luis Alvarez was put in charge of the project, which was located in Livermore (and, subsequently, became the site of the Lawrence Livermore National Laboratory). The accelerator was completed, but never used for its original purpose as extensive uranium deposits were discovered in the US. (See sidebar on MTA.)

In 1953, just after Donald Glaser (see sidebar in Chapter VII) had invented the first, very small bubble chamber, Alvarez undertook to develop liquid hydrogen bubble chambers of significant size for particle physics, culminating in the gigantic 72-inch chamber. With this chamber he and his group discovered many features of strong interaction physics. He received the Nobel Prize for this work. Bubble chambers, based upon Alvarez's success, were built in many laboratories throughout the world.

Some of Alvarez's other exploits, an outgrowth of his fertile imagination, which never ceased to be active, included "x-raying" the pyramids using cosmic rays to see whether or not there were hidden chambers, as well as a careful analysis of Zapruder's film to understand completely the events surrounding President Kennedy's assassination. The analysis was far better than that done by the FBI. He also invented a way to stabilize cameras, a method used in many commercial applications such as photographing the America's Cup races; as well as a device for simplifying, and making more accurate, execution of eyeglass prescriptions by opticians. He formed a private company to develop and sell these two inventions.

After he retired, Alvarez, working with his son Walter, Professor of Geology at the University of California, explored the extinction of the dinosaurs and showed that there was an asteroid that hit the earth just at the Cretaceous-Tertiary boundary (by the iridium found in those earth layers and brought to earth by the asteroid). This was the first time that it was shown that extra-terrestrial events would affect life on earth. It was an earth-shaking discovery that has changed, forever, man's view of himself.

Fig. 3.2 An accelerating tank of the first Alvarez linac, built just after WWII. Since that time many similar linacs have been built all around the world.

Fig. 3.3 The Materials Testing Accelerator (MTA), built in the early 1950s at a site that would later become the Lawrence Livermore Laboratory. The camera is looking down the vacuum tank of Mark 1. In order to be able to install the massive drift tubes, the team laid a standard gauge railroad track down the middle of the accelerator.

The machine produced a beam with great difficulty, mainly vaporizing a lot of copper bus-work, and never producing any fissile fuels. The discovery of uranium deposits in Utah in the early 1950s rendered it prematurely obsolete, and by 1952 it had been scrapped with hardly a trace remaining.

Alvarez's work on his first linac was taken up at the University of Minnesota by John Williams, who, in the 1940s and 1950s, constructed a 68-MeV linac. When strong focusing arrived on the scene in the early 1950s, John Blewett realized that quadrupole focusing would greatly improve linacs. Magnetic quadrupole lenses could provide transverse focusing and the grids were no longer needed. The quadrupoles could be mounted inside the drift tubes themselves. The first linac to employ quadrupoles in this way was the HILAC (see below).

Many proton linacs were constructed in the years that followed Alvarez's early work. These accelerators were mainly used as injectors into synchrotrons; first at the Bevatron in Berkeley and then at many other machines around the world. The Bevatron injector, a 6 m linac, produced 9.9 MeV protons and still had grids to produce radial focusing. This was a typical example of a linac designed as an injector for a proton synchrotron. Protons emerged from an ion source inside the negative high voltage terminal of a Cockcroft Walton set. They gained energy as they passed down an accelerating column to ground potential. They then entered the linac which, in later projects like the AGS and CERN PS, accelerated the particles to about 50 MeV. The higher the injection energy, the higher the current that could be injected into the synchrotron (see sidebar for space charge). The injection energy of still later proton injectors was typically 200 MeV.

The largest of these proton linacs was constructed by Louis Rosen at Los Alamos, not to inject into a synchrotron, but to produce a beam to study the physics of mesons. It was called the Los Alamos Meson Physics Facility (LAMPF). Construction started in 1968 and was completed in 1972. It was 800 m long and reached 800 MeV. It consisted of an Alvarez (or DTL) structure for the first 100 MeV, after which it switched to a structure more suitable for faster-moving particles, the Side Coupled Cavity (SCC) structure. Los Alamos pioneered the development of these and other very efficient configurations for the drift tubes of proton linacs and the resonant structure around them.

A very important idea, based on the work on linacs, was the invention of the radio-frequency quadrupole

Space Charge

The first accelerators were designed without regard to the effect of one particle upon another. One could call it the "independent particle approximation" that considers only the effects of external fields (magnetic or electric) upon the particles to be accelerated. That approximation was adequate for designing the early machines.

Soon, however, given the ability to achieve particle energies of interest, the desire developed to have ever more intense beams to improve the accuracy and quality of results. In intense beams, the interaction between accelerated particles becomes significant and the equilibrium between the external forces and the Coulomb interaction between particles — called "space charge" — becomes important. A simple but widely used "handy formula" to characterize this phenomenon was developed by L. Jackson Laslett in the early 1960s.

This formula gives the amount of change in transverse oscillation frequency of a particle due to its interactions with all the other particles. Clearly, that change must be kept within reasonable bounds and the Laslett formula is widely employed, even today, to estimate machine performance.

In the late 1950s accelerator designers began to contemplate colliding beams. Here, the requirement is to produce beams of such intensity that beam-beam collisions, between oppositely directed particles hitting each other, occur at an interesting rate. This also requires a very careful examination of the limiting effects: space charge effects are now to the fore. Two groups, one in the US and one in the Soviet Union, independently realized that not only would space charge lead to a static equilibrium, but that space charge could also lead to dynamical effects, called "coherent instabilities." In this mode the beam can undergo collective motion as the forces due to the beam's charge distribution act on the whole beam: twisting, turning, bunching it. These effects often would occur at a lower beam current than that given by the Laslett formula. There are many kinds of these collective instabilities, each with a colorful name that characterizes them, such as "negative mass," "head-tail," "resistive wall," and "electron cloud." Sadly there is not enough space here to explain these names. Various methods, including Landau damping, feedback, and careful design of the accelerating tube, can be used to control such instabilities. Without a detailed understanding of the phenomena, essentially no modern accelerator would operate anywhere near the intensity that they now reach.

The ultimate limit on intensity of colliding beams turns out to be an incoherent beam-beam tune shift very similar to the Laslett formula, but applied to two counter-moving beams. This appears to be an intrinsic limit, and although it has been studied for many decades, no one has devised a way to circumvent it.

The high-speed computers currently available allow one to simulate the effect of one particle upon another, and this method of attack is currently widely employed.

Radio Frequency Quadrupole (RFQ)

Invented in the Soviet Union by Teplyakov and Kapachinskii in 1970, the Radio Frequency Quadrupole linac (RFQ) was brought to the attention of Western physicists by Joe Manca at Los Alamos. The first RFQ, a "proof of principle" device built at Los Alamos, was small but highly successful. The RFQ is a low-velocity, high-current linear accelerator with high capture efficiency that can be employed to accelerate any species of nucleons from a few keV per nucleon up to a few MeV.

Instead of drift tubes, four vanes with appropriately spaced radial modulations (Fig. 3.4) are placed along the beam axis, and the cavity is excited in an electrical quadrupole mode. Strong transverse focusing comes naturally, while the longitudinal components of the electric field, from the radial vane modulations, provide bunching and acceleration of the beam. Because of the very fine periodicity possible in the modulations along the vane (each period corresponding to the effective length of a drift tube), it is especially suited to slowly moving particles. It soon replaced the very large electrostatic injectors (750 keV air-insulated Cockcroft-Waltons) and pre-accelerators found at older proton synchrotrons and thus have had a very large impact on the nature and cost of accelerators for protons and ions. RFQs as standalone accelerators are also finding many applications in ion implantation, isotope production, and neutron sources.

RFQs have been made of copper, copper-plated mild steel and aluminum as well as some other materials. Modern precision machining is employed on brazed or electro-formed monolithic structures in order to obtain the very high precision which is required.

More than 100 RFQs have been constructed. About a third are heavy ion devices; the rest accelerate protons. In the future they will undoubtedly be used in radioactive species accelerators, tandem electrostatic accelerator afterburners, and in many other devices requiring powerful beam sources.

(RFQ). This is a linac structure in which the drift tubes have a four-vaned cross section providing acceleration and focusing at the same time. It is specially suited to slowly moving particles and soon replaced all the rather massive 750 keV Cockcroft-Walton injectors that were the first acceleration stage in the injector chains of synchrotrons. RFQs were invented in 1970 by I.M. Kapchinskii and V.A. Teplyakov and promoted by Los Alamos (see sidebar for RFQ and Fig. 3.4).

There are two other applications of proton linacs which have been proposed for advanced physics: the Two-Beam Accelerator and the driver accelerators for Heavy Ion Fusion. They will be covered below in the section on induction accelerators.

III.4 Electron Linacs

Ising's original aim was to use his linac for electrons,

Fig. 3.4 The inside of a Radio Frequency Quadrupole. The RFQ has generally replaced the very large Cockcroft-Waltons (see Fig. 1.3) as the first stage of injectors into synchrotrons. The very complicated structure must all be machined using computer controlled machine tools.

but it took even longer for the necessary technology to become available for electron linacs than for proton linacs. Practical electron linacs were first developed at Stanford University. Stanford remained the leader in electron linacs up until very recently, when superconducting technology, again pioneered at Stanford but highly developed at other places, became important for electron linacs.

As we pointed out earlier, linacs for relativistic particles had all started with klystrons. (See sidebar for klystrons and Fig. 3.5.) The device itself had been invented by the Varian brothers prior to World War II and then, in 1945, Marvin Chodorow and Edward Ginzton invented high power klystrons. (See sidebar for Ginzton.) These klystrons typically operated at frequencies between 1 and 3 GHz. This was very much higher than the 200 MHz radar power tubes developed during World War II and used by Alvarez in his first proton linac. These high frequencies were called L-band and S-band and corresponded to wavelengths of 30 cm and 10 cm respectively. Magnetrons were used by William W. Hansen to make Stanford's first electron linac (see sidebar for Hansen), though all later machines used klystrons. At the time, Alvarez, who had already made his own proton linac, was not convinced by Hansen's plans and, according to Panofsky, even went as far as telling Hansen that it would be impossible. An electron linac takes advantage of the fact that electrons, even of low energy, are moving close to the speed of light. For example, a 0.5 MeV electron, easily produced by a simple electron gun, is traveling at 0.866 of the speed of light. At a somewhat higher energy, electrons will be so close to the velocity of light that acceleration will barely affect the distance they travel in one cycle of the RF wave, and the drift tube spacing need only change by a small amount along the linac. Rather than use a series of tubes suspended from the walls, the structure can be a conducting tube with periodic diaphragms. The diaphragms are like washers attached to the inner wall of the tube; their central hole allows the beam to pass. The diaphragms fulfill the function of the drift tubes of an Alvarez structure, but one may ask how the particle is protected from the adverse phase of the accelerating field. The answer is that the dimensions of the washers are chosen to ensure that the accelerating wave travels down the structure in phase with the particle. At a given point along the tube, one sees the electromagnetic wave go by, with the field direction ever changing from acceleration to de-acceleration; but the electron, riding along with the wave, is continuously accelerated.

It is possible to match the speed of the wave to the electrons because the diaphragms modify the velocity of

Klystrons

Powerful klystrons are at the heart of all electron linacs and electron storage rings. The story of their development would fill a fascinating book. The klystron was developed by the Varian Brothers before World War II and then made into powerful power sources after World War II by Edward Ginzton (see sidebar).

The Varian brothers, Russell (1898–1959) and Sigurd (1900–1961) were born of a long line of unorthodox Irishmen and women. Their parents were very poor and after moving a great many times settled in a small town on the coast of California. Sigurd engaged in barnstorming and exhibition flying and was a pilot for Pan Am in the Mexico and Central America region, in the days when there were repeated crash landings and ground conditions were marginal: snakes, earthquakes, political unrest, primitive conditions and natural hazards. Sigurd suffered from tuberculosis and quit Pan Am in 1935 but he became interested in helping pilots "see at night." He was also concerned about air attack of the US from Mexico and for both these reasons became interested in radar and the development of a power source. He engaged the interest of his brother, who was at the family home. His brother, after attending Stanford University by dint of walking and hitch-hiking from his home, had just been rejected by the Stanford graduate program. The two brothers convinced Stanford to let them use a lab in exchange for a share of royalties. The two of them employed the idea of Hansen's rumbatron (see sidebar for Hansen) along with the key idea of bunching of a stream of electrons, invented by Russell in 1937, to produce the first microwave power source, which they called a klystron.

This power source was further developed by the Sperry Gyroscope Company from 1937 until the end of World War II. The Varians, Hansen and two graduate students, one of whom, John Woodyard, later became professor at the University of California in Berkeley; all were at Sperry on Long Island. The klystron did provide a safe landing in a test in Boston, but was not chosen as the main power source of radar. The magnetron (developed in England) turned out to be the device used in all the radars of World War II as the main power source. There was however a low-power klystron in the radar units.

After the war, the scientists returned to Stanford and leased some land, becoming the first members of what later became the very famous Stanford Industrial Park. This was the start of Silicon Valley. There, due to the efforts of Ginzton, the klystron was turned into a very powerful device, making possible the electron linacs of Stanford and many other places around the world.

Fig. 3.5 The insides of two klystrons. One can see how carefully they are designed and how complicated are the beam structures. When surrounded with various components to supply power, remove power, cool the structure, etc. they are often larger than a person.

Edward L. Ginzton (1915–1998)

American physicist
Member of the National Academy of Engineering
Member of the National Academy of Sciences
Morris Lieberman Memorial Prize, 1957
IEEE Medal of Honor, 1969

Ginzton was born in Russia but became a refugee from the Russian Revolution when he was 13 years old. In 1929 he was able to emigrate to California, via China. Although he could not speak a word of English when he arrived, he went on to obtain a BS in 1936 from the University of California in Berkeley, and in 1937 an MS. He then moved to Stanford where he obtained an engineering degree in 1938 and a PhD in physics in 1940.

At Stanford he met William Hansen (see sidebar) and the Varian brothers (see sidebar on klystrons). In 1941, the four of them moved to Sperry Gyroscope, where they all worked on the klystron, and where Ginzton rapidly advanced to become head of microwave research.

In 1946 Ginzton returned to Stanford as Professor of Applied Physics and Electrical Engineering and led a team that developed the high-power klystron. Teaming up with Hansen, the two developed the first electron linear accelerators. As early as 1950, and stimulated by Henry Kaplan, he began to work on the medical application of such electron linacs.

In 1956 he was the leader of a group of scientists and engineers that were studying the feasibility of a two-mile accelerator, then called Project M (for "monster"). In 1959 he was asked to become Chairman of the Board of Varian Associates and in 1961, just after the death of Sigurd Varian, and knowing that Project M was well on the way to being approved, he left Stanford and devoted himself to full time work at Varian.

He loved skiing, hiking and sailing. He was an avid restorer of Model A Fords and a consummate photographer. He never sold his pictures although they were of professional quality. He had four children, all of whom partook in the family hikes, skiing holidays, and outings.

William W. Hansen (1909–1949)

American physicist
Member of the National Academy of Sciences

Hansen (right in this picture) was raised in Fresno, California, where his father owned a hardware store. He was intensely interested in gadgets, electrical equipment, and mathematics; and at an early age decided he would become an engineer. He was very precocious, having graduated from high school at the tender age of 14, and after having to wait a year, entered Stanford, to study engineering at the age of only 16.

In 1928 he worked as a laboratory assistant in the physics department, developed a love for physics, and became a graduate student in the subject. After only his first year he was made an instructor in physics. In 1933 he received a PhD and, with a National Research Fellowship, he spent some time at the University of Michigan and more time at MIT.

At MIT he studied under P.M. Morse and became interested in electromagnetic fields and how they are affected by boundaries: a subject to become of vital importance a few years later to him for his invention of the "rumbatron."

In 1934 Hansen returned to Stanford as an Assistant Professor of Physics, and remained at Stanford, except for the war years, until his very untimely death, at age 40, from respiratory distress. The disease that killed him had caused shortness of breath for many years but was made worse by a very strenuous regime during World War II and particularly aggravated by a trip to England in 1948, for an accelerator conference, which Hansen had forced himself to attend.

In the late 30s Hansen invented the "rumbatron," which now seems so obvious and is taught in even elementary courses in electricity and magnetism, but at the time was a revolutionary advance. It was this idea that led to the first klystrons, and later, made electron linacs possible.

In 1941 Hansen and his whole group moved to the Sperry Company where they worked on radar. In 1945, Hansen returned to Stanford as a full professor of physics. In the remaining years he was at the spearhead of efforts to develop an electron linac. It is said that even Luis Alvarez didn't believe this effort would be successful, and expressed this opinion to Hansen — fortunately without much effect.

Hansen was known as a warm human being, most generous with both advice and equipment. He was shy and quiet in social gatherings, but fond of jokes. He was a strong individualist and a champion of human rights.

the electromagnetic accelerating wave as it passes down the tube, producing a "slow wave." Since even a 50 MeV electron is going at 0.99995 times the speed of light, the change in velocity needed is very small; but the structure must be "just right" in order to have the slow wave move at exactly the correct velocity. The machining of the accelerating tube is done as precisely as possible, but the ultimate fine tuning is made by taking a C-clamp and dimpling each 3 cm section, indenting it ever so slightly.

Stanford produced a number of electron linacs during the 1950s, among them the Mark III, which accelerated electrons to 1 GeV. Robert Hofstadter, who was awarded the Nobel Prize in 1991, used electrons from this machine for his scattering experiments to reveal nuclear properties. At MIT, an electron linac of 17 MeV was constructed in 1951 and followed by a larger machine, 400 MeV, proposed in 1964, funded in 1966, and completed at the Bates Linear Accelerator Center in 1973. The really large electron accelerator built at SLAC, close to Stanford University, was proposed under the direction of E.L. Ginzton, but very soon became the responsibility of W.K.H. Panofsky. The project was at first called Project M (for Monster) and, after many years of planning, fund raising, and pacifying the neighbors (1960–1965), emerged as a two-mile accelerator that accelerated its first 20 GeV electron beam in 1966 (see Fig. 3.6). The project was led by Richard Neal; key physicists in the construction of the two-mile accelerator were Greg Loew and Roger Miller (see sidebars for Loew and Miller). The accelerator was built with a bridge over it, destined later

to carry a major freeway. This machine, the centerpiece of the new national laboratory, the Stanford Linear Accelerator Center (SLAC) (see sidebar for SLAC) opened up a rich vein of physics research, including the Nobel Prize-winning work (1990) that in 1968 revealed the parton (quark) structure of protons and neutrons. The accelerator was used in due course to provide particles for the storage rings SPEAR, PEP and PEP-II, as described in Chapter VI. In the 1980s, after installing more powerful klystrons to raise the energy to 50 GeV, it was used for the first linear collider: the Stanford Linear Collider (SLC). Currently the 2-mile accelerator is being prepared to drive an FEL, the Linac Coherent Light Source (LCLS) described in Chapter VIII.

Once technology caught up, the idea of a linac began to forge ahead. Superconducting RF cavities were found to offer considerable advantages for linac designers. Room temperature cavities dissipate large amounts of power in their walls, at the high field levels needed to accelerate particles. It was realized that if one

---

W.K.H. Panofsky (1919–)

American physicist
Honorary degrees from many universities in six different countries
Atomic Energy Commission E.O. Lawrence Award
California Scientist of the Year Award
Enrico Fermi Award of the U.S. Government
National Medal of Science
Foreign Member, Chinese Academy of Sciences, Académie française, Russian Academy of Sciences
Member of the National Academy of Sciences

W.K.H. Panofsky, called by everyone who knows him, Pief was born in Berlin and entered the United States in 1934. He obtained his BA from Princeton University in 1938 and his PhD from California Institute of Technology in 1942. Both Pief and his brother were at Princeton at the same time. Both were, naturally, at the top of their classes, but one was just a bit better than the other. In order to be clear about which person was meant, they became known as "The smart Panofsky" and "The dumb Panofsky."

Pief's talents were recognized early, and during World War II, 1942–43 he was made head of a military research project at Caltech. He participated at Los Alamos in diagnostic work on nuclear weapons. In 1945 he went to the University of California in Berkeley, quickly moved through the academic ranks, and remained there until imposition, by the State of California, of a Loyalty Oath for faculty members, to which he objected. Thus in 1951 he moved to Stanford University, where he remained.

In Berkeley Pief worked closely with Luis Alvarez (see sidebar). At Stanford he was Director of the High Energy Physics Laboratory from 1953 to 1961 and then the Director of the Stanford Linear Accelerator Center (SLAC) from its inception in 1961 until his retirement in 1984. Without Pief there would have been no SLAC, and he set the direction and organized the research that has made that Laboratory world-famous. In fact, a number of members of SLAC have won Nobel Prizes for work that was only possible thanks to Pief. He has had a major influence on high-energy physics, not only through his own research and, indirectly, through the impact of SLAC, but also through his advisory capacity to the Atomic Energy Commission and then the Department of Energy. Pief's views were listened to over and over again for his counsel was invariably wise and sensible. He has served on dozens of important high-energy physics advisory committees — it is difficult to think of one on which he did not serve and surely his views, opinions, and judgments were sought by the few that he was unable to attend.

Pief has broad expertise beyond that in high-energy physics. He has been an advisor on such wide-ranging matters as the detection of nuclear explosives, arms control, nuclear materials control, and post cold-war deterrence policy. He has served on many very important committees in this broader field, such as the President's Scientific Advisory Committee (1960–64), the Arms Control and Disarmament Agency (1959–80), and the General Advisory Committee on Arms Control to the President (1978–80).

Pief has worked tirelessly on matters of arms control. In the National Academy of Sciences he was a member, and then chair (1985–93), of the Committee on International Security and Arms Control. Under his direction this committee developed useful interactions with the Soviet Union, and then China, during periods when US relations were minimal.

Pief is married to Adele DuMond (the daughter of a famous physicist) and they have five children.

Fig. 3.6 The SLAC site, just west of the main Stanford University Campus with its two-mile long linear accelerator. The two arms of the SLC linear collider are indicated, converging on the experimental area, as is the large ring of PEPII fed in both clockwise and counterclockwise directions. The damping rings at the front end of the linac are used to reduce the emittance of the beams for the collider and maximize luminosity.

Gregory A. Loew (1930–)

American physicist
Fellow of the American Physical Society
Recipient of the French Medal of the Ordre du Merite

Gregory Loew was born in Vienna, Austria in 1930, and raised in Paris, France and Buenos Aires, Argentina. He received his Licence-ès-Sciences degree from the Faculté des Sciences at the Sorbonne in Paris in 1952, his M.S. in Electrical Engineering at Caltech in 1954, and his PhD in Electrical Engineering at Stanford University in 1958. His entire career was spent at Stanford, starting in 1958 with Project M (for "Monster"), later in 1962 to be renamed SLAC. He has held a variety of positions at the Laboratory, beginning in 1964 as head of the Accelerator Physics Department, then Deputy Director of the Technical Division (1980–2001), and finally, Deputy Director of SLAC (2001–2005). He was and still is a Professor at SLAC (which has an academic affiliation with Stanford University), having started as an Adjunct Professor in 1974, and moved up to Professor by 1982. As an avocation, he created and then taught a course at Stanford on the "Causes of War" for over ten years.

From SLAC's inception, Loew was involved in the design and construction of the linear accelerator, working on accelerator structures, radiofrequency systems, instrumentation and beam physics. He participated at various levels in the development of all the other current and future machines of the laboratory, SPEAR, PEP, SLC, NLC, and LCLS, and took a special interest in the behavior of accelerator structures at very high electric gradient. He was also instrumental in the creation of the Continuous Electron Beam Accelerator Facility, CEBAF, now the Thomas Jefferson National Accelerator Laboratory, having proposed the initially accepted design (an Electron Pulse Stretcher Ring), which was later superseded by the current superconducting machine. His other activities at SLAC include the procurement and management of electrical power for the lab and the administration of most of the international collaborations. When the Superconducting Super Collider, SSC, was terminated, he participated in the gigantic (but sad) disposition of leftover equipment (1993–1995).

Loew has played an important role in advancing the status of large electron-positron linear colliders. He chaired both International Linear Collider Technical Review Committees, the first in 1994–1995, the second in 2001–2003. This work in 2004 eventually culminated in the selection of a preferred technology for the International Linear Collider.

Loew has been active on human rights matters for many years and was chair of the American Physical Society Committee on the International Freedom of Scientists in 1996. His interests, other than physics, include international affairs and politics. He loves music and has a special fondness for opera.

Roger Heering Miller (1931–)

American physicist

Born in Ohio and with a BA from Princeton University in 1953 and a PhD in physics from Stanford University in 1965, Miller has spent his entire professional life at Stanford. In earlier time, from 1953 to 1956, he was a Damage Control Officer and then an Engineering Officer on a destroyer in the US Navy. He held various research posts from 1959 to 1961 in the Microwave laboratory at Stanford University before joining the Stanford University Linear Accelerator Center (SLAC) where he remained until his retirement in 1997.

At SLAC he rose to become Professor of Applied Research from 1983–1997, a position he retains in an Emeritus role.

During his years at SLAC he made many contributions to linear accelerators, electron beam optics, wake-field suppression in accelerator structures, injection, and positron sources. He has been a quiet worker who simply gets things done. His theoretical and experimental ability has been outstanding.

His ability is widely appreciated, for he has served in many different capacities as advisor and has been a consultant to Haimsen Research, SAIC, Brobeck Inc., and Beta Development Corp.

Stanford Linear Accelerator Center (SLAC)

The Stanford Linear Accelerator Center (SLAC) started in 1962 under the leadership of W.K.H. Panofsky (see sidebar). When Project M, as it was initially known, located near the football stadium on the Stanford Campus, was approved, a contract was signed, and construction started on the two-mile accelerator. There had been concerns in Congress about the idea — so much so that Varian had been asked to agree that it would not build klystrons for the project. There were also objections from neighbors who did not like the visual impact of its power lines. Eventually a pleasant design was developed for such utilities that satisfied all; this design has subsequently been employed in many other locations. By 1966 construction was completed and research commenced. The design energy of 20 GeV was achieved the following year.

Many important discoveries in particle physics were made with this powerful linac. There was the discovery of the point structure within a proton in 1968. With this came the realization that quarks, which up until then had been a purely theoretical concept, were actually real. This work rewarded the three investigators with a Nobel Prize (1970).

Next came the electron storage ring, SPEAR. It was built on a shoe-string (not as a genuine construction project) and was completed in 1972. "SPEAR" stands for Stanford-positron-electron-asymmetric-ring, but the asymmetry was too expensive and the ring had to be built in a symmetric configuration — the acronym remained. In 1974 this ring was the site of the "November Revolution" in which the psi particle was discovered. At the same time it was independently discovered at Brookhaven National Laboratory and called the "J" and since then always been known as the J/psi. A shared Nobel Prize was awarded for this only two years later (1976).

The muon is just like an electron, only heavier. No one understands why this is the case. On SPEAR, in 1976, a new meson, the tau, again just like the electron, but heavier even than the muon, was discovered. Thus there are three electron-like particles, called "leptons" — and there are reasons to believe there are no more. This important discovery was awarded with the Nobel Prize in 1982.

In the 70s another larger electron storage ring, called PEP, was constructed in collaboration with the Lawrence Berkeley Laboratory and completed in 1980. "PEP" stands for proton-electron-positron; the ring was only funded for electrons and positrons, but the acronym has remained. (In Germany a much bigger ring colliding electrons with protons was subsequently built.) In 1994 construction was initiated, again with Berkeley as a participant, on an asymmetric electron-positron storage ring, using much of PEP and called PEP-II. The asymmetry allows study of the B meson, which has a particularly large amount of time-reversal non-invariance, which is today an especially important subject of study in high-energy physics. Experimental study using PEP-II was initiated in 1999. A similar B factory has also been built in Japan.

In the 80s, SLAC built two arcs at the end of the two-mile accelerator, and with various gymnastics was able to study the interaction between the two beams. This device, the Stanford Linear Collider (SLC) was the first linear collider. It came into operation in 1989, with 50 GeV electrons on 50 GeV positrons (a significantly higher energy than the original 20 GeV of the SLAC accelerator), but was almost immediately in competition, as far as physics output is concerned, with the much more intense LEP ring at CERN. In recent years, SLAC has devoted considerable resources, as have many other laboratories around the world, to the necessary R&D for a high-energy linear collider.

All the developments described above are in high-energy physics — the primary purpose of SLAC. At the same time the use of SPEAR as a synchrotron radiation source of x-rays was being developed. In 1990 SPEAR had its own injector, so it was functionally independent from the two-mile accelerator. Although the synchrotron radiation use, called the Stanford Synchrotron Project (SSP) and then Stanford Synchrotron Radiation Laboratory (SSRL), started in 1973, it was not until 1992 that SSRL became a Division of SLAC. Most recently, in 2002, SLAC has initiated work on a linear accelerator free-electron laser, called the Linac Coherent Light Source (LCLS). This device promises coherent x-rays, at 0.15 nm and with a very short pulse, as is described in Chapter VIII.

could use superconducting materials to line the cavities, the power dissipated would be much reduced and the average number of electrons accelerated could be much improved. This development has become particularly important in the context of future linear colliders whose performance (see sidebar on luminosity in Chapter VI) would benefit greatly from a continuous beam. The use of superconductivity for the RF cavities of an electron linac had been pioneered during the period from 1964 to 1981 by Harry Alan Schwettman and Todd Smith on the Stanford University campus (close by, but distinct from, SLAC). Operation of this machine was not easy, but has had an impact on the performance of nearly all recent and future electron linac projects.

Early superconducting accelerators were put to a number of purposes: most notably for powering the Stanford Free Electron Laser (FEL), a concept developed at Stanford by John Madey (see sidebars on John Madey and on FELs in Chapter VIII). Research and development of superconducting RF cavities has more recently been pursued at Cornell and at CERN for LEP to achieve the highest accelerating fields in circular machines (see Chapter VI) — advances which were rapidly adopted by the linac community. In 1976 it was decided that the world needed a long pulse electron linac to study the internal nature of the hadrons. In the spirit of fairness a competition was held in the US, between MIT, the University of Illinois, the Argonne National Laboratory, Southeastern Universities Research Association (SURA), and the National Bureau of Standards. SURA was selected, the Newport News site was chosen, and Hermann Grunder (see sidebar for Grunder) appointed as Director.

The Argonne proposal had been for a superconducting linac but SURA's was not. However, once selected, SURA adopted superconductivity for its continuous electron beam accelerator facility, called CEBAF. The laboratory bore this name until, in 1996, it was officially named the Jefferson Laboratory; the name CEBAF is now reserved for the accelerating facility itself. This is a novel machine in which the beam is recirculated through the same two linacs five times. The superconducting linacs are like the straight parts of a race track, while four quite distinct semicircular arcs of magnets at each end — each with its strength matched to a different energy — recirculate the beam through the linacs. After five complete passes around the track, each time through a different arc, the beam reaches 4 GeV. This ingenious design is a compromise between using the same two expensive linacs as many times as possible and being able to design a stack of arcs

Hermann A. Grunder (1931–)

Swiss-American physicist
U.S. Senior Scientist Award by the Alexander von Humboldt Foundation, 1979
Director of the Jefferson Laboratory, 1985–2000
Scientist of the Year for 1998 in the Commonwealth of Virginia
Distinguished Associate Award by the U.S. Department of Energy, 1996
Honorary Doctorate, University of Frankfurt, 2000
Director of the Argonne National Laboratory, 2000–

Born in Basel, Switzerland, he first obtained a mechanical engineering degree from the University of Karlsruhe in 1958, and then, in 1967, a PhD in experimental nuclear physics from the University of Basel, Switzerland. Although this may seem to be a long time to spend crossing the Rhine, he had in the meantime been far from inactive and had been very busy broadening his horizons in the US. He was first in the Mechanical Engineering Department at the Lawrence Berkeley Laboratory from 1959 to 1962 and then, from 1962 to 1964, associated with the commissioning of Berkeley's spiral ridge cyclotron. He did experiments for a year with this machine before returning to Switzerland to complete his education.

In 1968 he rejoined the Lawrence Berkeley Laboratory, this time in the Accelerator Division. In 1979 he was made an Associate Director and Head of the Accelerator and Fusion Research Division. His most important achievement during this period was the conception and construction of the Bevalac — a combination of a heavy ion linac and the Bevatron, which pioneered the fields of relativistic heavy-ion physics, and radiation therapy with light-ion beams (described in Chapter IX).

In 1985, Grunder became the Director of a new laboratory, now called the Jefferson Laboratory, where a machine was built to provide a continuous beam of electrons for nuclear physics studies. A competition had been held to design this machine, and although the new laboratory "won" the competition with their own scheme, Grunder completely changed the design for a superconducting RF accelerator. The machine was built under his supervision and has served the community admirably.

In 2000, Hermann Grunder was selected to be director of the Argonne National Laboratory, where he used his skills not only to direct the laboratory, but to foster collaboration among federal research facilities, universities and industry to further the cause of science. He stepped down as director in 2005.

Among the important national committees that Grunder has served, many have had a significant affect on the physical sciences. He has led international efforts to gain acceptance for safe, proliferation-free nuclear energy. He is a keen supporter of the biosciences, and a strong advocate for the role of technology in homeland security. He is also the proud recipient of no fewer than eight honorary degrees.

In his free time he enjoys the outdoors: hiking, biking, and skiing.

of magnets at each end. The facility started operation in 1994.

In Europe, the DESY Group has pursued development of a very efficient, very high gradient, superconducting accelerator structure. Their work, in close collaboration with Cornell, has shown the way to ever-higher RF gradients, which lead to shorter accelerators. The DESY Group proposed the superconducting version of the International Linear Collider (ILC), called TESLA, whose concept is now at the heart of the ILC. In Chapter X we will explain why this mode, rather than a room temperature design, has been selected for the ILC. While the world finds a site, and the necessary support, for the ILC, DESY is going ahead testing a full scale section of TESLA as a driver for an FEL.

III.5 Heavy Ion Linacs — a Rich Field of Research

We have charted the history of proton and electron linacs but must now return to 1955 to describe another important kind of linac — the heavy ion linac. We have already mentioned the first of these machines: the HILAC, which Berkeley constructed in the years from 1955 to 1957. The design was prepared together with Yale University and a similar machine was built at Yale. The accelerator could handle ions with light atomic weights and most of the experiments used ion species in the atomic mass range of carbon and oxygen. After all the careful design of the HILAC had been completed, the cyclotron pioneer E.O. Lawrence (see sidebar in Chapter II) still wanted to see a cyclotron instead of the linear machine for heavy ions. He called a meeting in support of this view and the participants, perhaps selected for this purpose, had only good things to say about cyclotrons. However, Lawrence was flexible and receptive of new ideas and, having listened to the arguments, became convinced that a linac was indeed better, withdrew his opposition and remained very supportive even when the machine went through the usual difficult "running-in" period.

The Berkeley Group then proposed a versatile machine, the Omnitron — a synchrotron rather than a linac. This was years ahead of its time, and would have accelerated all atomic species to high energies. It was not funded, perhaps due to the lack of strong support by the Laboratory Director, Ed McMillan, or perhaps because of Berkeley's opposition at that time to the Vietnam War. In 1971–1972, following the demise of the Omnitron, the HILAC was significantly upgraded to become the SuperHILAC. This new machine, 49 m long, accelerated ions up to xenon (with an atomic weight of 136) over a range of energies from 1.2 to 8.5 MeV/nucleon. Both machines used John Blewett's idea of quadrupoles for transverse focusing instead of the grids at the drift tube ends (see sidebar for Blewett). HILAC was, in fact, the first heavy ion linac to have such quadrupoles. In 1981 a third injector was added to the machine so that it could accelerate ions of all atomic numbers up to uranium.

A substantial program in transuranic physics was undertaken at Berkeley with the HILAC, SuperHILAC, and other machines. Many super-heavy elements (lawrencium, californium, berkelium, americium, seaborgium, etc.; twelve in all) were discovered — in fact, all elements up to atomic number 106 except for 105. The discovery of transuranic elements that began in Berkeley, spread to GSI in Germany and to Russia.

The Gesellschaft für Schwerionenforschung (GSI) had been constructed at Darmstadt in Germany during the period of 1966–1969. It consisted of a Wideroe linac followed by two Alvarez type linacs for a total length of 74 m. This UNILAC achieved its design specifications in 1975 and accelerated all species up to the atomic number of uranium, at energy of 3 GeV (or 12.6 MeV/nucleon). Here too, new elements, such as bohrium, hassium, darmstadtium, and meitnerium were discovered.

Many super-heavy ions have been discovered at the Joint Institute for Nuclear Research (JINR) at Dubna, where they used tandem cyclotrons rather than linear accelerators. They are credited with discovering the transuranic element, atomic number 105, dubnium and the — yet to be named — elements 113–116.

The Japanese were not to be left out of this rich field of research and in the period of 1975–1979 a linac was built at the Institute of Physical and Chemical Research (RIKEN), which could accelerate atomic masses up to 160 MeV for krypton and 180 MeV for xenon.

Also during this time a large number of other laboratories constructed heavy-ion linacs, many as energy-increasing "afterburners" for Tandem accelerators.

The focus for high energy ions began to shift to circular machines which still used linacs as injectors. In about 1971, Albert Ghiorso (the person who has discovered more elements than any other — 12 at last count) realized that the SuperHILAC could be connected to the Bevatron and thus one could have very energetic ions of all species. This prompted Edward McMillan (see sidebar in Chapter II), then director of the lab, to remark, "Why didn't I think of that?" The Transfer Line connecting these two machines was completed in 1974 under the leadership of Herman Grunder (before he left Berkeley for Jefferson Laboratory) and, after pioneering

John Paul Blewett (1910–2000)

Canadian/American physicist
American Physical Society R.R. Wilson Prize, 1993
Founder of the journal Particle Accelerators

Blewett, born in Toronto, earned his bachelor's and master's degrees in physics at the University of Toronto and his PhD in physics from Princeton University in 1936. After receiving his doctorate, he spent a year at the Cavendish Laboratory in Cambridge, England, working under Ernest Rutherford and Mark Oliphant, on, among other projects, range-energy relations for alpha particles.

From 1937 to 1946, Blewett worked in the research laboratory of the General Electric Co. in Schenectady, New York. During this time, Blewett came across a paper by two Russian theoretical physicists in which they pointed out that high-energy electron beams circulating in a betatron would lose some energy by radiation. Blewett concluded that the radiation would indeed be significant and would make it difficult to build machines for higher energy. He predicted that the radiation would cause the orbit to shrink in the GE betatron (100 MeV) and performing an experiment with that machine, observed the shrinkage — the first evidence of what is now known as synchrotron radiation.

During a visit to the Radiation Laboratory at the University of California, Berkeley in 1945, Blewett learned of McMillan's ideas for a synchrotron. Blewett and colleagues decided to build what they hoped would be the first operating synchrotron. However, the first to demonstrate the principle were neither McMillan or Blewett, but rather Frank Goward and D.E. Barnes in England.

In 1947 Blewett actually missed the first visual observation of synchrotron radiation from the new GE electron machine. By that time he had left GE, along with his wife Hildred Blewett, who was also an accelerator physicist, to join the new Brookhaven National Laboratory.

After six months of being "security risks," they had obtained the necessary clearance to join the BNL staff, where the first task was to work on the construction of a new particle accelerator, the Cosmotron, which was to extend the energy of accelerated particles by a leap of an order of magnitude to 3 GeV. Blewett took charge of the design and construction of the magnet and the radio-frequency accelerating system. In 1952, the Cosmotron came into operation as the world's first billion-volt machine.

When Courant, Livingston and Snyder, as is described in Chapter V.7, came up with the invention of strong focusing, Blewett immediately realized that this concept could also be put to great advantage in linear accelerators. In particular, replacing the grids in linacs with quadrupoles was a huge improvement.

When the new European high-energy physics laboratory, CERN, was proposed in 1952, a group of physicists from several European countries visited BNL where they learned of the new idea of strong focusing. This group invited John and Hildred Blewett to join them to help with the new laboratory. They went to Bergen, Norway, where they contributed to the design of the CERN Proton Synchrotron (PS), soon to be built at the new site in Geneva.

In early 1954 he moved back to BNL where the Alternating Gradient Synchrotron had been approved for construction. Ken Green (see sidebar in Chapter V) was in charge and Blewett his deputy. In 1960, a 33-GeV proton beam was achieved and Blewett soon moved on to become involved with Luke Yuan in a study of the possibilities of proton synchrotrons up to 1000 GeV.

In 1962, Blewett, together with Stanley Livingston of MIT, published the book *Particle Accelerators* (McGraw-Hill, 1962), which summarized the development in the field up to that time, serving as a standard text for many a year. In 1970 he founded and became first editor of a new journal of the same name.

After 1978 and during his "retirement," he returned to an early interest: synchrotron radiation. He took part in initiating the proposal to build the National Synchrotron Light Source at BNL. He was also a consultant to the Synchrotron Radiation Research Center in Taiwan.

Blewett was a bon-vivant, who enjoyed fine restaurants, as well as the outdoors through sailing and gardening. He shared these many activities with his second wife Joan, who for a very long time was associate director of the History Center at the American Institute of Physics.

work in two important fields, was finally turned off in 1992. The first of these was the use of ions in cancer therapy (as described in Chapter IX), and the second was the study of high-energy nuclear physics with ions. The idea of taking heavy ions from a linac and adapting a synchrotron to further accelerate them soon revealed important nuclear physics results. The idea caught on, and experiments were scheduled on a series of machines of ever increasing energy. The Alternating Gradient Synchrotron (AGS) at Brookhaven was modified for ion physics and then the 28 GeV Proton Synchrotron (PS) at CERN was given a new injector and ion linac to send ions on their way to the 400 GeV SPS. Subsequently a

dedicated ion collider, RHIC at Brookhaven, was built in the search for a quark-gluon plasma (see Chapter VI).

A major advance in the development of ion accelerators was the use of superconducting RF, pioneered by the Argonne Laboratory for a machine called the Argonne Tandem Linear Accelerator (ATLAS). The heart of ATLAS is the superconducting split-ring resonator (see Fig. 3.7). The first successful test of a niobium split-ring resonator occurred in November 1977 after two and a half years of intensive development. Niobium was chosen as the superconducting surface because it is the best available superconductor for a radio-frequency device. First acceleration of ions was in 1978 and, with

Fig. 3.7 A picture of the world's first superconducting heavy ion accelerator, ATLAS, built during the 1970s at the Argonne National Laboratory in Illinois. The picture shows the innards of the accelerator: the accelerating structure with its drift tubes supported on helical arms.

its new Positive Ion Injector (PII) added in 1992, ATLAS can now provide beams of all ions up to uranium.

The nuclear physics community, having tapped such a rich vein of research, would now like to see an accelerator capable of accelerating unstable isotopes as their next major construction. Such a device would greatly extend our knowledge in nuclear physics. At GSI a project with this very goal, FAIR, is under construction. In the USA there are plans to construct the Rare Isotope Accelerator (RIA). This machine would probably have a superconducting linac as its main component. These projects will be described in Chapter X. Similar or even more ambitions projects are under discussion in Japan and in China.

III.6 Induction Linacs

The reader will now be somewhat mesmerized by the series of linacs we have described — none of them very different from the periodic cavity structures proposed by Ising more than 80 years ago. Induction linacs are quite different. In the next chapter, on betatrons, we will be introduced to the idea that particles may be accelerated in a circular path without any radio frequency cavities, but simply by the electro-motive force generated around the circle by changes in the magnetic flux that links their circuit. We can think of this as an application of Faraday's law in which the accelerated beam behaves as the secondary of a transformer. We can change the betatron transformer's topology somewhat so that the beam passes along the axis of a simple ring of iron or, more usually, ferrite. A primary winding also links the ring; a pulse of current in this winding changes the flux in the ring so that an accelerating force acts on the particles in the direction of the ring's axis. An induction accelerator consists of a series of these rings spaced along its length rather like the drift tubes of a more conventional linac. Radio frequency linacs produce long trains of pulses, each of modest peak current of the order of 100 milliamps. In contrast, an induction accelerator is a device that produces large currents (as much as 10,000 A) at a low repetition rate (usually at about one Hz, but more recently up to many kHz).

Despite a very successful history of invention, design and construction, induction accelerators, unlike many of the devices we describe in these pages, are not numerous. However, there are a number of possible applications that may result in induction linacs being more widely used.

It was N.C. Christofilos (see sidebar in Chapter V) who first developed an induction accelerator, namely the Astron injector, in the late 1950s and 1960s. His motivation was to build an accelerator that could supply a cylindrical layer of electrons to a fusion device. This layer (the e-layer) was to provide stability to the fusing ions.

In the Astron, charge stored in a capacitor was released, producing a pulse. The pulse then passed through a primary circuit and was transformed into an accelerating field — the beam of charged particles was the secondary of this "transformer." In the Astron the capacitance consisted of standard coaxial cables charged up to 25 kV. Thyratrons switched the discharge current into the Ni-Fe cores, which acted as the transformer yoke. The energy of the accelerated beam was a modest 3.5 MeV but a giant beam current of up to 350 A lasted for 300 nanoseconds. The repetition rate of Astron was 60 Hz. There were more than 300 switches in the 10 meter length of the accelerator. It was completed in 1963 and then, in 1968, upgraded to 6.0 MeV and 800 A (see Fig. 3.8). It was finally closed down when the emphasis in fusion research switched to tokamaks.

Fig. 3.8 The world's first induction accelerator, Astron, built at the University of California Lawrence Radiation Laboratory (later to be called the Livermore Laboratory) in the late 1950s and early 1960s. One can see the very massive iron "cores," which in modern machines are replaced with much more compact ferrite cores fed by rapidly switching pulse-forming networks.

Astron was put to another use in the late 1960s when a novel idea, the electron ring accelerator (ERA or "Smokatron"), was under study at Dubna and then at Berkeley. This was a "collective" accelerator, which means the accelerating force came from forces associated with other particles surrounding, preceding or following the bunch of beam to be accelerated. The concept, the invention of V.I. Veksler (see sidebar in Chapter II), required the formation of an intense ring of electrons, which collected protons by ionization of the background gas. The ring, and its proton load could be accelerated along its axis by allowing it to expand in a specially shaped solenoidal field. The idea was that the light electrons could be accelerated rather easily, and in a short distance would be very close to the velocity of light. The protons, trapped within the ring, would then have the same velocity as the electrons but, being much heavier, would have acquired a much higher energy. To make the electron ring in the first place required an induction accelerator.

The first experiments of the Berkeley Group were on the Astron injector. Initial positive results prompted

them to develop a better injector. The new ERA injector produced 1 kA of 4 MeV electrons with a pulse length of 45 ns and a repetition rate of 5 Hz. This accelerator had ferrite tile cores, 17 spark gaps, and oil-filled storage lines at 250 kV for energy storage. The innovations were used in all the subsequent induction accelerators. The ERA program was abandoned in Berkeley, but carried forward at Garching and Dubna to the point of accelerating some protons. The concept turned out to be correct but it was very cumbersome, very difficult to accomplish in practice, and had a disappointing acceleration gradient. There is perhaps a cautionary tale here — Europe was at one time under pressure to abandon plans to build its 300 (later 400) GeV SPS because "the ERA would be much more compact, simpler and of course cheaper." Fortunately the pressure was resisted.

In addition to the Astron and the ERA Injector, two powerful induction accelerators were built at Livermore, under a program to develop particle beam weapons. The Experimental Test Accelerator (ETA) was completed in 1979 and the Advanced Test Accelerator (ATA) in 1983. The ETA produced 10,000 A of electrons with a pulse length of 30 ns, at energy of 4.5 MeV, with a repetition rate of 2 Hz. The ATA was much larger, 50 m long, and together with its power supplies filled a very large building. It produced 10,000 A of electrons of 50 MeV, with a pulse rate of 5 Hz. The weapons program for which it was intended was abandoned in the mid 1980s. Although primarily funded as a weapon, the ETA did serve a useful purpose as an injector to a very successful FEL (described in Chapter VIII) used in a basic science program. This produced 1.8 GW of power, at 30 GHz, and first demonstrated the technique of Self-Amplified Spontaneous Emission (SASE) for an FEL (see Chapter VIII for an explanation of SASE and related new ventures). The success of the FEL at ETA stimulated "Star Wars" activity involving the ATA. The goal was a ground based FEL for shooting down satellites and incoming missiles. The FEL on the ATA did not live up to its promise and the program was terminated.

Another very successful application of induction accelerators, also associated with weaponry, has been to develop intense pulses of x-rays for studying the implosion process of nuclear weapons. The FXR at Livermore was located at their remote "site 300" and, when completed in 1982, produced 3,000 A of 18 MeV electrons (see Fig. 3.9). This machine, beautifully built and, like ETA and ATA, based upon the ERA induction linac technology, has been very useful for flash radiography. In the early 1990s FXR was followed by the Dual-Axis Radiographic Hydrotest Facility, DARHT: a machine constructed at Los Alamos for the same purpose. Under the Stockpile Stewardship Program it has become important to check if nuclear weapons are no longer cylindrically symmetric. This might be due to a rust spot developing on one side of the weapon, or simply because it had been stored on one side for decades. A second axis to DARHT has therefore been constructed to find such asymmetries and this facility was run-in during the first years of this century. (See Fig. 3.10.)

III.7 Applications of Induction Linacs

We turn once again to peaceful uses of induction accelerators and to plans for the future powerful linear accelerators that will be used to collide electrons and positrons for the next generation of high energy experiments and those beyond as part of a future Linear

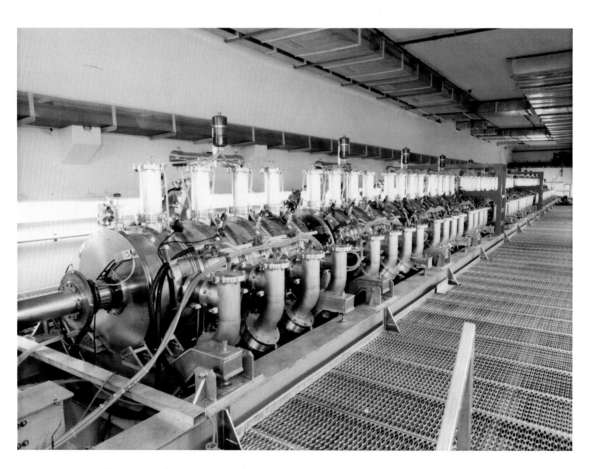

Fig. 3.9 A later induction accelerator, FXR, built, at Lawrence Livermore National Laboratory, in order to study the behavior of the implosion process in nuclear weapons. The facility was completed in 1982 and the contrast with the Astron is most noticeable.

Fig. 3.10 The Dual Axis Radiological Hydrodynamic Test Facility (DARHT) at the Los Alamos National Laboratory, New Mexico. This device is devoted, as was the FXR, to examining nuclear weapons from two axes rather than just one. This reveals departures from cylindrical symmetry, a sign of aging which can seriously affect performance.

Collider. We shall return in detail to linear colliders in Chapter X. For the moment we would like to highlight one of its more novel manifestations: the Two-Beam Accelerator in which an intense "drive beam" is used to produce power that is then used to accelerate the "main beam" to very high energy. Thus the hundreds of klystrons of a conventional high energy linac are replaced with a single or a small number of drive beams. The Berkeley group proposed an induction accelerator to produce the drive beam for such a scheme while the CERN group preferred an RF linac, assembling the intense beam by combining sections of a very long pulse train. The Berkeley group built an induction linac to study the behavior of the intense drive beam, but the program was terminated for lack of support despite scientifically encouraging initial results. The CERN study — the Compact Linear Collider or CLIC — continues; we shall describe this in some detail in Chapter X.

Another possible future application of induction accelerators is to help solve the world's energy crisis. Starting in the mid-1970s, there has been considerable interest in the production of fusion energy by means of inertial implosion of a fuel pellet (deuterium-tritium) following the impact of intense particle beams.

Fusion only occurs when the product of temperature and time for such an implosion reaches a critical value, the Lawson criterion. In an effort to reach these conditions, GSI in Germany pursued the course of making the required very intense particle beam by superimposing sections of a long pulse train from an RF linac into one short pulse. A Berkeley group has, characteristically, taken the induction accelerator approach. But despite the many significant advances that both groups have made in the last 30 years, the programs have been greatly reduced and may be near termination. More details are to be found in Chapter X.

There has also been recent work on the development of induction synchrotrons at KEK in Japan. The basic idea is simply to replace the RF cavities in a circular machine with induction cells. This now seems possible, in the context of some accelerators, thanks to advances in technology that allows very rapid pulsing. It is hoped that the collision rate in colliders can be greatly increased using this technique. Initial experiments have been very positive and the idea is now being considered at a number of laboratories.

Still another use of induction accelerators (or at least a very small induction unit) is in the cooling of beams.

Very low temperatures of ion beams can be made much more orderly, or given "low temperatures," by using an induction unit to sweep the beam energy across the resonance of a cooling laser.

Other possible uses of induction accelerators — none of which are as yet financially supported — include the powering of sub-critical nuclear power reactors. Such reactors are intrinsically safe as they quench immediately when the accelerator is turned off. The same reactor allows both the burning of nuclear wastes to remove the very long lifetime actinides and the breeding of natural uranium or natural thorium into a reactor fuel. Induction accelerators may also be used in the exploration of high energy density plasmas and the production of a neutron spallation source.

Chapter IV. Betatrons

The betatron is a circular device that accelerates electrons: beta rays as they were originally called in early observations of radioactivity. Following the success of the cyclotron it seemed that the betatron, and the synchrotron which was to come later, should also be called "-tron" to show that they too were accelerators.

Betatrons made a slow start. The cyclotron was the first practical accelerator for more that a few MeV and it accelerated heavy particles like protons and deuterons — but it would not work with electrons. We saw in Chapter II how even protons, at 30 MeV when they come close to the speed of light, no longer remain in step with the RF in a cyclotron. The electron, being almost 2000 times lighter than a proton, is, at any particular energy, much closer to the velocity of light and therefore loses synchronism at a few keV.

Electrons were not very interesting at that time; physics was concentrating on how neutrons and protons were bound together in the nucleus. A force, which we now call the "strong force," seemed to attract neutrons and protons, overcoming the repulsion of the electrical charges of the protons, which, left to themselves, would have made the nucleus fly apart. It seemed natural to use a particle that would exert this force — neutron or proton — as a probe and not the electron, which seemed only to have electrical properties.

Of course electrons had been used to produce x-rays from a copper target for medical diagnosis since the days of Roentgen, but the electron energy required was not very high. If you look at the knobs and dials on your dentist's x-ray machine you will see it is calibrated to produce up to 100 kV: a voltage quite within the range of an electrostatic generator. Later on there was a need for intense, higher energy electron beams to produce harder x-rays that would penetrate the body and treat tumors. In Chapter II we mentioned that at the Berkeley laboratory in the early 30s Sloan had used a resonant transformer to provide 800 keV electrons

for this purpose. This was not a betatron but a transformer whose secondary winding was an open circuit. Electrons experiencing a fraction of a wavelength of the oscillating high voltage in the gap were thrown off at high energy.

When the quantum theory predicted that atoms would emit and absorb x-rays and that these would reveal the energy levels of the atom, a controlled source of x-rays was required; but here again the energies needed to excite the K level electrons are only of the order of a few keV.

And so for many years there seemed no pressing need for high energy electron accelerators. Although betatrons, first discussed as early as 1922 (under the name of "beam transformer"), would have filled the gap, it was only in 1942 that the idea was taken up again to accelerate electrons in the energy range 5 MeV to 300 MeV. The popularity of the betatron was short-lived for, by 1948, electron synchrotrons had been developed which were more compact, lighter, and therefore less expensive. Synchrotrons could easily be built for a higher energy than the betatrons they superseded. Nowadays only a few betatrons are built — as a portable source of hard x-rays.

Those who built betatrons had to tackle and overcome a number of technical difficulties that were the keys to building other circular machines (synchrotons and colliders) which followed. Betatrons, though they had no accelerating cavities and used magnetic induction to accelerate electrons, were the first circular machines to have a rising field that was kept in step with the rising energy of a pulse of electrons. Unlike cyclotrons that produced a constant stream of beam, the electrons in a betatron were accelerated in a series of pulses — just as they are in the synchrotrons that followed. To keep the particles on a circular orbit as they were accelerated, designers had to understand how they would oscillate, from side to side and up and down about the ideal circular path. These oscillations, which are just as important for today's synchrotrons and storage rings, are still called "betatron oscillations," their frequency the "betatron frequency" and their magnitude the "betatron amplitude." If they get into resonance with field errors, that is called a "betatron resonance" — homage to the basic understanding contributed by the designers of the betatrons.

IV.1 Early History

The name "betatron" is relatively new and does not appear until Kerst revived the idea in the 1940s. Two decades earlier they were called "beam transformers" or "ray transformers." It was J. Slepian who, in 1922, had the idea of employing magnetic induction to accelerate electrons. He took his idea from the voltage transformer to be found in all common household appliances: the same device that powers the lap-top on which this book is being typed. In an electrical transformer, an alternating current in the primary coil generates an alternating magnetic field in a soft iron magnetic core. Both the primary and secondary coils are wound on the same magnetic core so that the changing field excites alternating current in the secondary.

Slepian's idea was simply to replace the secondary winding with a free stream of electrons flowing in a ring-shaped vacuum chamber, which, like the secondary of a normal transformer, was threaded by the core. The current that flows in a normal transformer's secondary is driven by the voltage induced in the coil from end to end. At each point in the coil there is an electric field (the gradient of this voltage) that drives the electrons along the copper wire of the coil. There would be a similar voltage, V, around any circular path around the magnetic core. Suppose the secondary coil were replaced by a donut-shaped vacuum chamber. Any free electrons in the donut would feel the corresponding electric field and start to circulate, gaining an energy, eV, on each turn. It was also argued that the device would be very efficient since the losses in a normal transformer's core are very small.

Slepian's first thoughts were patented in 1927 while he was working at the Westinghouse Electric Company and although his work was purely theoretical, it stimulated some sporadic experimental work over the next two decades. In his device, electrons were held in a circular orbit by the magnetic field between two poles of an electromagnet — just like in the cyclotron that was to follow. The electromagnet was in fact the core of the transformer and carried magnetic flux lines through the orbit of the electrons. It was just opened up enough to slide in the donut-shaped vacuum chamber. The primary winding of the transformer was the same coil that excited the electromagnet. Thus the field between the poles both guided the electrons in a circle and accelerated them. With some clever matching of flux and field the orbit radius would remain the same.

In the fall of 1922 and before any experimental work was initiated, Rolf Wideroe, while still an undergraduate, was having similar ideas. He called his device a "ray transformer" ("beta ray transformer" might have been a better description). In his device there were two coils, one to provide the magnetic field necessary to keep the electrons circulating at a fixed radius and the other to produce the flux which varies in time and accelerates the electrons. His principal contribution to the concept was to think carefully about the relation between the two fields. He found that in order for the guide field to increase at a rate which keeps pace with the required rate of acceleration, it should have exactly half the value you would expect in a simple dipole magnet carrying the same flux. This "Wideroe condition" is essential for a betatron.

Wideroe described the whole concept of a betatron in his notes in 1923, and applied for a patent. This would have predated Slepian's work but there is no record that Wideroe pursued his application to the end. In the autumn of 1925, Wideroe wanted to develop the concept as an experiment for his doctor's degree, but his thesis advisor encouraged him to look for another topic!

In 1926, Wideroe moved to Aachen, where he was able to start experimental work on his ray transformer. He built a device that was to reach an energy of 6.8 MeV but unfortunately it never worked, probably because electrons charged up the glass walls of the accelerating ring.

At the same time (1927) and far away in the Bureau of Terrestrial Magnetism, in Washington, Breit and Tuve were also experimenting with an induction device. They were motivated by the desire to make bursts of x-rays. Their machine did not satisfy the Wideroe condition, but they were content to let the electrons spiral inwardly as the magnetic field increased. However, they did realize the importance of focusing; shaping the field near the orbit such that, if an electron deviated slightly from its design trajectory, it would experience a force that would make it return to the correct trajectory. For vertical (axial) focusing this required a magnetic field that decreased with increasing radius. This feature was exactly the opposite of what had been foreseen in Slepian's design but it proved to be correct and has been incorporated into the design of all subsequent circular accelerators: cyclotrons, betatrons, synchrotrons, etc. The experimental work of Breit and Tuve probably resulted in a working machine, but the detection techniques for bursts of x-rays were so rudimentary that one does not really know. In any case, about this time, the success of electrostatic machines (described in Chapter I) brought the efforts on induction accelerators to an end.

Meanwhile in Britain, E.T.S. Walton was also pursuing the concept of an induction-based circular accelerator. His outstanding theoretical work verified the concept of the Wideroe condition and, more importantly, quantified the radial focusing condition that had been first appreciated by Breit and Tuve. He also did extensive experimental work, but had difficulty with the injection of the electrons. Like other workers, he had made no provisions for the extraction of electrons other than forcing them to hit a target and produce x-rays. His experimental work was not rewarded with success, but his theoretical contributions were important for all further circular machines.

A number of other ideas were put forward in this period. In 1936, Jassinsky proposed adding electrostatic fields to supply radial focusing, a concept that would only work at low energy. Also in 1937, M. Steenbeck, of the Siemens Company in Berlin, applied for a patent on a circular induction accelerator which was very similar to Wideroe's. Steenbeck claimed, subsequently, that in 1935 he was able to accelerate 1.8 MeV electrons although there is no indication in the patent that he realized the need for conducting walls, nor is there description of any experimental work. His experiments were not pursued.

In 1938 G.W. Penny applied for a patent for a device very similar to Wideroe's, but with electrostatic focusing. F.R. Abbott, at the University of Washington, built a device that had all the elements that one needed to make a betatron (including conducting walls), but for unknown reasons the device never worked and his efforts were never described in a publication.

Also never published was the work of James L. Tuck and Leo Szilard at Oxford University. Both of them carefully designed the machine, although it was Tuck who did the experimental work. The device was designed to reach an even higher energy than that of Breit and Tuve. The work of Tuck and Szilard was the most complete in technical detail and the most likely to succeed of all the various attempts we have described. However, the advent of World War II brought the work to a halt, and the device was never operated.

IV.2 The Kerst Betatron

In 1939, D. Kerst at the University of Illinois extended Walton's work on the theoretical motion of electrons in spatially varying and time dependent fields. This analysis, due to Kerst and Robert Serber, paved the way for the first successful betatron and provided a solid basis for the analysis of particle motion in all future circular accelerators. Serber was a first-class theorist and J. Robert Oppenheimer's right-hand man in Berkeley in the 30s. Later he became a theorist at Los Alamos, where he is famous for developing a handbook, which every newcomer read, that described the key physics of atomic bombs. However, the basic ideas were all due to Kerst, and Serber's role was "only" to elegantly formulate the considerations.

Kerst was not to be put off by the previous unsuccessful attempts to build betatrons (see sidebar for Kerst). He persevered and on July 15, 1940 operated the first successful betatron. (See Fig. 4.1.) This was a circular tube of radius 7.5 cm, producing electrons with an energy of 2.3 MeV. The initial x-rays from this device were equivalent to 100 millicuries of radium; later this was increased to the equivalent of a few grams of radium. One can imagine the impact that this development made in medicine as well as in nondestructive material testing (for example, looking for flaws in welds). Kerst then took a year's leave from the University of Illinois in 1940–41 and went to the General Electric Laboratories where he designed two betatrons: a 20 MeV machine and a 100 MeV machine. (See Fig. 4.2.) During the World War II days, Kerst built a 4 MeV portable betatron for inspecting unexploded bombs in the field. He was also at Los Alamos where, according to the official history, his "technical achievements are amongst the most impressive (at the laboratory)." Among these was a 20 MeV betatron that was used to study the dynamics of bomb assembly implosions.

IV.3 The Wideroe Betatron — Second Attempt

Reports of Donald Kerst's success reached Wideroe in Trondheim in the autumn of 1941. Wideroe immediately returned to his earlier ideas on the "ray-transformer." He wrote a rather extensive review paper that was published in 1943.

In the spring of that year and while living in Nazi Germany, he was asked to go to Berlin and work for their war effort. At that time his brother had been sentenced to 10 years in jail for helping refugees escape

Donald William Kerst (1911–1993)

American physicist
Comstock Prize of the National Academy of Sciences, 1945
Elected a Member of the National Academy of Sciences, 1951
James Clerk Maxwell Prize of the American Physical Society, 1984
American Physical Society R.R. Wilson Prize, 1988
Honorary degrees from four institutions

Donald Kerst was born in Illinois and earned his BA and PhD from the University of Wisconsin in 1934 and 1937. He was at the University of Illinois from 1938 to 1957, and concurrently spent extended periods at the General Electric Company and Los Alamos. In 1957 he moved permanently to the General Atomics Laboratory and then in 1962 to the University of Wisconsin, where he remained until he retired.

In his early years he developed the betatron, which we describe at length in Chapter IV. He built a series of ever larger version of these devices, at Illinois, General Electric, Los Alamos, and again at Illinois. From 1953–57 he was the technical director of the Midwestern Universities Research Association (see sidebar on MURA). At General Atomics he started a whole new career in plasma physics pursuing the search for fusion energy, a quest in which he made many important contributions.

His love of physics was legendary and infectious. He had a wonderful way with youngsters, treating them as equals, giving them direction, but letting them exercise imagination and often go off in unexpected but innovative directions. He almost always insisted that their names should be listed first among the authors of papers and they should be the ones to make the presentations at meetings. During his lifetime he had more than 70 students.

He kept a list on his blackboard of various problems, ideas, and directions, and he was forever hoping "to put the last nail in the coffin" of some uninteresting or incorrect idea. Of course this never happened and the list just grew longer and longer. Kerst had a good sense of humor; for example, he often remarked, as some bad effect was uncovered, that he was concerned that this bad effect would not become so important that it would be named after him.

He was devoted to his family and loved the outdoors. Even in his later years he was able to paddle a canoe on the lakes near Madison, Wisconsin, far more powerfully than many men much younger than himself. He was delighted when metal skis first became available; he bought himself a pair, and as he put it, "It allows one to ski the way he did 25 years ago." He also enjoyed sailing and deep sea fishing.

Fig. 4.1 One of the first betatrons built in the early 1940s: the 20-inch machine at the University of Illinois.

Fig. 4.2 Standing in front of the 100 MeV betatron are E.E. Charlton and W.F. Westendorp, who completed the machine (in the early 1940s) at the G.E. Research Laboratory in Schenectady after Kerst had returned to the University of Illinois. J.P. Blewett describes it as *"a splendid object emitting 2600 Roentgens per minute and a deafening noise when operated at 60 cycles per second."*

from the Nazis and there were hints that his brother's term would be shortened. The Nazis didn't disclose their purpose to Wideroe until 1944, but it emerged that they were interested in using the x-rays from a betatron as a long-range death-ray to kill enemy pilots or detonate their bombs. There was little chance that this would succeed.

Meanwhile work started on what was known as the Hamburg Betatron. This 15-MeV machine, financed by the German Aviation Ministry, was completed in 1944 and produced x-rays equivalent to about 30 grams of radium. A much larger 200 MeV machine was designed but never built. Wideroe's brother's prison term was not shortened and he had to wait to the end of the war when he was liberated by the Americans. After World War II, Wideroe was himself imprisoned, this time by his own country, Norway, for collaboration with the Nazis. The Norwegians incorrectly suspected that he had invented and worked on the V2 rockets. He was released once they realized that the betatron work had not resulted in any practical weapon.

IV.4 The Years After World War II

After the war, Kerst returned to the University of Illinois and there, in 1950, after building an 80-MeV machine, he completed the largest betatron ever built — a 300-MeV machine — used for the study of particle physics.

At this time there was considerable industrial interest in betatrons, both for medical purposes and for non-destructive material testing. Their commercial production was in the hands of General Electric, Westinghouse, Allis-Chalmers, Brown Boveri, Siemens-Reiniger and Philips. Their machines were in the range of 30 MeV and many were sold as a safer alternative to an intense radium source. Their obvious advantages were that they could be turned off when not required, and that the beam could be directed to a particular spot.

The use of these machines to provide electrons and hence x-rays for the radiation treatment of cancer patients pioneered the widespread use of accelerators in medicine. In Chapter IX we will read that cyclotrons had been used for medical treatment even prior to World War II, but their use was very limited and always involved bringing the patient to a physics laboratory. More than 200 betatrons were built after World War II and installed in hospitals in many locations in more than 20 countries. Nowadays, of course, these betatrons have given way to linear machines. The new linacs are cheaper, more reliable and easier to operate.

Perhaps the last time we hear of a betatron in accelerator research was when one was converted by Goward and Barnes into the first "proof of principle" synchrotron (see Chapter V). Nowadays, except for those still produced as portable x-ray sources (see Fig. 4.3), betatrons are only of historical interest.

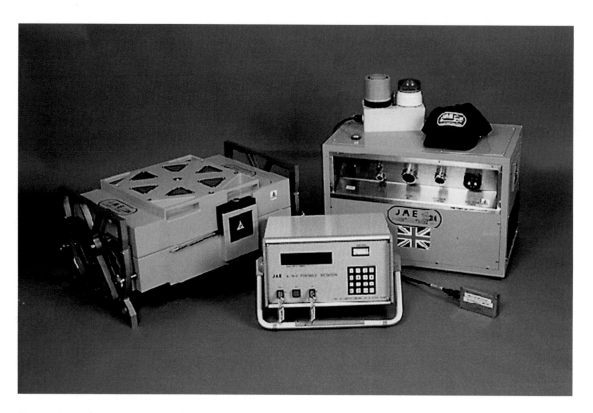

Fig. 4.3 A modern, very compact betatron, commercially produced, and with selectable output energy up to 6 MeV. It is used to produce x-rays to look for defects in large forgings, castings, valves, steel beams, ship's hulls, pressure vessels, engine blocks, billets, thick welds, dense metals, composites, military hardware, munitions propellants, reinforced concrete buildings and bridges.

Chapter V. Synchrotrons

V.1 Science Motivation

Although the cyclotrons and betatrons described in Chapters II and IV were able to apply a modest accelerating voltage many thousands of times, their size grew with their output energy. Their magnets had pole pieces that saturated at about 2 T and the only way to double the top energy of the particle, for example, was to make the circular path twice as big. If the poles were twice as big, the volume of steel in the magnet yoke would be eight times larger. Lawrence's largest cyclotron already had as much steel in it as a naval frigate and it looked as if a whole fleet would have to be sacrificed to make a significant leap in energy. It seemed to be the end of the road. But then it was realised that only the outer rim of the magnet was needed to channel particles at the top energy and, if particles could be persuaded to use this rim at all energies, there was a hope that the huge cyclotron might be replaced by a slender necklace of magnets around this rim. To make this work, particles would have to be injected into the machine in pulses; each pulse would then be accelerated and the magnetic field of the necklace increased to keep in step with the rising energy. Physicists had to wait until World War II was over before they had the time to invent and fully develop this idea, but when they did they called it the synchrotron. The name comes from the need to synchronise the rise in field — and, as it turns out, also the accelerating frequency — to match the rising energy of the bunch of particles as they are accelerated.

The synchrotron has proved a trustworthy companion in the quest for higher energies. The ongoing development of the synchrotron principle from machines of a few hundred MeV to monsters of several TeV was made possible by two important discoveries, that of alternating-gradient focusing and the use of colliding beams. The world's two most recent big synchrotrons, both at the European laboratory CERN near Geneva, are LEP for electrons and the LHC project for protons. These are many kilometres in girth, but recent years have also seen a diversification of powerful lower energy machines for the production of synchrotron radiation, production of neutron beams by spallation, and for proton and light ion therapy.

V.2 The Early History of the Synchrotron

At the beginning of the Second World War, the skills of cyclotron builders in the US were diverted to the task of the electromagnetic separation of uranium. Colleagues from the UK joined them in this and, although some later regretted their role, the stimulating company, coupled with an ample time to reflect, proved an ideal environment for creating new ideas. In 1943 Marcus Oliphant (see sidebar for Oliphant), an Australian physicist who had been working in England at Birmingham University, found himself at Oak Ridge supervising the business of transforming a laboratory experiment for isotope separation into a large scale industrial process. As deputy to E.O. Lawrence he was often given the owl watch and "with little to do unless troubles developed" occupied his time by speculating on plans for his return to Birmingham when the war was over. He wrote a memo to the Directorate of Atomic Energy, UK in which he proposed a new method of acceleration — the synchrotron. This contained the essence of the idea:

> "Particles should be constrained to move in a circle of constant radius thus enabling the use of an annular ring of magnetic field... which would be varied in such a way that the radius of curvature remains constant as the particles gain energy through successive accelerations by an alternating electric field applied between coaxial hollow electrodes."

His new idea was not greeted with enthusiasm at a time when more important business was afoot, but, as he left for England at the end of the war, he was encouraged by Lawrence to pursue the idea further.

After his return to Birmingham, Oliphant read the comprehensive and beautiful papers by McMillan (a student of Nobel prize-winning Lawrence) and Veksler (a Soviet physicist). McMillan described Oliphant's pulsed ring-magnet idea and announced his own plan to build such a machine — without a single reference to Oliphant.

V.3 First Synchrotron

McMillan's publication of plans to build a synchrotron was a challenge to others to get there before him. Late in World War II the Woolwich Arsenal Research Laboratory in the UK had bought a betatron to x-ray unexploded bombs in the streets of London, of which there were many at the time. Frank Goward, a physicist at Woolwich, read of the synchrotron idea and realized that he could become the first person to make it work by converting the betatron into a synchrotron with the help of a rudimentary accelerating electrode. The influence of Lord Rutherford's "string and sealing wax" spirit of improvisation on a generation of UK physicists is perhaps to be seen in Fig. 5.1, which shows this historic machine, with which they just scooped their American rivals.

V.4 Electron Synchrotrons

A team constructing the first purpose-built synchrotron at the General Electric Co. at Schenectady (the betatron factory) just failed to beat the Woolwich people to the post by a month or two but they had the consolation that their 300 MeV machine, with a glass vacuum chamber, was the first to produce a new phenomenon — synchrotron radiation in a visible form. It had long been predicted that deflecting electrons in a magnetic field would emit electromagnetic waves, but no one knew what their frequency would be. Witnesses report that they had searched with a radio receiver scanning through the long and medium bands. Imagine the surprise when an intense stream of visible light emerged from the glass donut that held the electron beam in this early synchrotron. The synchrotron light is clearly to be seen in Fig. 5.2.

V.5 Early Proton Synchrotrons

The first synchrotrons had been electron machines but projects for proton synchrotrons aiming at energies

Marcus Laurence Elwin Oliphant (1901–2000)

Australian physicist
Inventor of the synchrotron
Pioneered centimeter radar and the atomic bomb

Mark Oliphant was an impressive individual, tall, with thick white hair. He spoke convincingly and energetically and in later life described himself as "a belligerent pacifist." Educated at Adelaide University, he joined Ernest Rutherford's group at the Cavendish Laboratory, Cambridge in 1927. He and Rutherford were the first to identify tritium and helium-3 produced by bombarding light nuclei with protons and deuterons.

By 1937 he was a Professor at Birmingham and, charged with overseeing Britain's coastal defenses, saw the clear need to use a radar wavelength much shorter than the 170 cm then current. His department set about developing a suitable power source and before long John Randall and Harry Boot from his department invented the magnetron. This was a device that made it possible for aircraft to carry their own radar and upon which the fate of the war hinged. He took this precious gift to the USA in 1941 where there were the resources to develop and mass produce it.

About this time he had been convinced of the practicality of an atomic weapon using uranium enriched by electromagnetic separation. He instigated the MAUD Committee in the UK, whose positive findings on feasibility of the bomb were at first ignored when transmitted to the US, where they remained "locked in a safe." On a visit to the US he enlisted Lawrence's help in bringing the idea to the attention of influential people who initiated the Manhattan Project, which eventually led to the production of the two bombs dropped on Japan.

Although he joined this project himself, he was later appalled by the consequences and returned to Birmingham at the end of the war vowing never again to have anything to do with the military. He became a founding member of the Pugwash Movement, attending many of their meetings.

Back in Australia he pursued his idea of a 10 GeV synchrotron powered by a homopolar generator. Unfortunately he was not able to master the ambitious technology that involved liquid sodium electrical contacts. When he retired he moved into public life as the governor of South Australia. He was open-minded and became popular with the public during his five-year term of office starting in 1972.

above 1 GeV were not far behind. Oliphant was back at the University of Birmingham in the shadow of the replica of the Siena campanile that adorns that campus. He had made his bid early to construct a 1 GeV proton machine but he was becoming bogged down in the red tape and lack of imagination that abounded in post-war Britain. Europe was not yet used to big science — after all, the Manhattan Project had been on US soil. Much

Fig. 5.1 Late in World War II the Woolwich Arsenal Research Laboratory in the UK had bought a betatron to x-ray unexploded bombs in the streets of London. Frank Goward converted the betatron into the first "proof of principle" synchrotron, whose completion just preceded their American rivals.

Fig. 5.2 This 300 MeV electron synchrotron at the General Electric Co. at Schenectady, built in the late 1940s, was the first purpose-built synchrotron. The photograph shows a beam of synchrotron radiation — the first to be observed — emerging from its glass vacuum chamber.

of his workforce had to come from the graduate students in his department.

There were technical problems to be solved too. Synchrotrons follow a pulsed "heartbeat." Particles are injected at low energy and then accelerated over several seconds before being deflected onto a target at high energy. Acceleration in a synchrotron is provided not by the ring of bending magnets, which merely keep the particles in a circular path, but by a copper cavity driven by a radio transmitter. This must be tuned to follow the huge swing in revolution frequency as the particles velocity around the ring increases. Such tuning had never been tackled before and Oliphant's solution involved plunging a large coil of copper wire, the inductor of a resonant circuit, into a mercury bath. Filing the shape of a rotating disc — the capacitor of this resonant circuit — proved an effective if irreversible means to make a fine adjustment.

The strength of the electromagnets that guide the beam must also match the momentum of the beam as it is accelerated. The pulsed current for the magnet

excitation for the Birmingham machine was produced via a direct current generator driven by an alternating current motor. This output of the generator was servo controlled by light reflected from 120 mirrors on the periphery of the radio frequency capacitor.

Oliphant's machine reached just short of 1 GeV for the first time in July 1953 — a few months later the Cosmotron at Brookhaven reached 3 GeV and closely followed by the 6 GeV Bevatron at Berkeley, which started up in 1954. Figure 5.3 shows the completed synchrotron in Birmingham.

The race between the Berkeley and Brookhaven laboratories was keenly contested. The Cosmotron team, which included Stan Livingston, John and Hildred Blewett, Ernest Courant, Ken Green and N. Blackburn, came in first. It was in May 1952 that *The New York Times* headlined their first "Billion Volt Shot" (see Fig. 5.4).

Early synchrotrons relied only upon the weak focusing fields produced by tapering the gap between the poles of the guide magnets towards the inside. There

Fig. 5.3 Although the first proton synchrotron to be planned, this 1 GeV machine at Birmingham University lagged behind the Cosmotron and Bevatron and achieved its design goal only in 1953. Its designer, Oliphant, had meager resources and had to use mainly graduate students to stack the laminations and wind the coils of the huge C-shaped magnet.

Fig. 5.4 The 3 GeV Cosmotron was the first proton synchrotron to be brought into operation. The Brookhaven team built it in competition with rivals at Berkeley, California, building the 6 GeV Bevatron. In those days calculating how big the donut vacuum chamber had to be was largely a matter of guesswork. Brookhaven was optimistic and reduced their aperture to a mere 1.2 by 0.22 m, which made it an easier and faster machine to build. When they first switched on in May 1952, *The New York Times* acclaimed their first "Billion Volt Shot."

Fig. 5.5 Overview of the Berkeley Bevatron during its construction in the early 1950s. One can just see the man on the left.

was a lot of controversy about the size of the gap necessary to allow for the mis-steering of the injected beam and the effect of magnet imperfections. It was difficult to estimate regardless of whether one was an optimist or a pessimist. A fall-back solution for the Bevatron had a huge magnet aperture, 4.3 × 1.2 m, which gave rise to rumors that it was destined to be the world's most powerful accelerator of Jeeps (which we are told could drive between the poles, provided their windshield was down). The Cosmotron had boldly refined their aperture to a mere 1.2 × 0.22 m. Apart from such matters the Bevatron had other troubles, as its builders had been distracted for a year or two in mid project to construct a large accelerator as part of a defense project (see Section III.4). Fig. 5.5 shows the Bevatron as it was finally constructed. The construction team was led by Edward Lofgren.

V.6 Nimrod and Phasotron

The UK decided to follow the Birmingham machine with an 8 GeV proton synchrotron. Nimrod had a huge magnet aperture and was powered by a motor generator-alternator whose load rose from zero to 100,000 horsepower in 0.75 seconds. The energy to be used for the magnetic field, was stored in a huge flywheel which was driven from the mains and which speeded up and slowed down as the magnet pulsed. One night a pole on the rotor of the alternator broke and it was only the heroism of an operator racing towards the circuit breaker over a catwalk above the monster, already writhing in its death throes, that prevented the export of a large rotating flywheel across the Channel to France.

The history of the synchrotron in the Soviet Union followed lines parallel to that in the West. Their first application of the phase stability principle, discovered by Veksler and independently by McMillan, was in the synchrocyclotron or "Phasotron" as Dubna called it. The first operation of this machine was timed to be on Stalin's 70th birthday, which we are told was a Soviet tradition. Dubna's 10 GeV weak focusing "Synchro-phasotron" (see Fig. 5.6) surpassed the 6 GeV Bevatron and, from 1957 until the 25 GeV Proton Synchrotron at CERN was finished, offered the highest energy in the world.

V.7 Strong Focusing

The Cosmotron, a typical weak focusing synchrotron, had a "C" shaped magnet open to the outside. Any imperfections in the magnetic fields or mis-steering of

Edward Lofgren (1914–)

American physicist
Fellow of the American Association for the Advancement of Science
Fellow of the American Physical Society

Born in Chicago, he received his undergraduate education at the University of California in Berkeley where he was an Assistant in Lawrence's Radiation Laboratory from 1938–1940. He was a group leader at Los Alamos during the war years. While working there he completed his post-graduate work for the University of California which granted him a PhD just after the war. From 1946–48 he was at the University of Minnesota and during this time consolidated an ongoing association with the Radiation Laboratory, which would become the Lawrence Berkeley National Laboratory (LBNL).

He was involved with essentially every accelerator project at the LBNL until his retirement in 1982 and particularly in the construction of the largest accelerator at LBNL: the Bevatron. This machine aimed to be the first accelerator to get into the GeV range (then called BeV) and produced some important physics, including a number of Nobel prizes. He became the group leader for the Bevatron after its completion in 1954.

Lofgren was an Associate Director of LBNL from 1973 to 1979. It was under his direction that the SuperHILAC was built, connecting the linear heavy ion accelerator with the circular Bevatron. (The construction of the Bevalac was led by Herman Grunder — see sidebar in Chapter III.) This machine ran one-third of the time for biology and two-thirds for nuclear physics. Pioneering work on cancer therapy with ions was carried out which provided the basis for the many machines later built specially for therapy. Nuclear physics at HILAC also provided the basis for further work with ions at CERN and for the construction of the RHIC machine at Brookhaven National Laboratory. A program in heavy ion fusion was also initiated under Lofgren's direction.

In the early 60s there was a move to build a much bigger proton synchrotron than the Bevatron. The initial target was 200 BeV. Lofgren led the study team at Lawrence Berkeley that produced a report that was used as a basis for costing and site selection for the new project. Although Lofgren would no doubt have preferred the machine to be built near LBL, a site for what was called initially the National Accelerator Laboratory and later Fermilab was chosen close to Chicago. The task of constructing this new ring was placed under the direction of Robert Wilson (see sidebar). Even though the final design turned out to be rather different than Lofgren's, the LBL design was instrumental in getting approval for funding of Fermilab.

Lofgren served in a number of advisory capacities, most importantly as a member of the High Energy Physics Advisory Panel (HEPAP) from 1967–1970.

the beam entering the accelerator would send it heading off to hit the sides of the vacuum chamber before it could be accelerated. There had to be some magnetic

Fig. 5.6 Overview of the Dubna Synchro-Phasotron, for a brief time the highest energy machine in the world. When its existence was disclosed by Veksler, in a private meeting in 1955, it caused consternation in the United States (remember this was at the height of the Cold War) and resulted in the authorization of the Zero Gradient Synchrotron at the Argonne National Laboratory. The ZGS was to be built on as fast a time scale as possible and with an energy just greater than that of the Dubna machine. In fact, the Bookhaven AGS tripled the energy of the Dubna machine and operated only a few months after ground was broken for the Argonne machine in 1959.

focusing in both the horizontal and vertical direction. To provide this focusing, the magnets were shaped to create a field gradient with a radius that should be negative but not too strong. As the magnet was excited, the outer parts of the poles tended to saturate and this made the field there weaker and the gradient stronger — too strong for horizontal stability. The upper energy of the Cosmotron was limited by this effect. Stan Livingston (see sidebar in Chapter II) had the idea of compensating saturation by reinstalling some of the C magnets with their return yokes (normally to the inside of the ring) towards the outside, but he was worried in case the focusing at low energy would be affected.

This led, almost by accident, to a much better way of focusing the beam thanks to Courant, Livingston, Snyder and Christofilos — and thereby lies a story. Courant had been given the task of checking the effect of alternating the yokes of the Cosmotron and reported that — far from being harmful — the focusing seemed to improve as the strength of the alternating component of the gradient increased. Snyder, as befits a good

theorist, who should always be ready with an *a posteriori* explanation, reminded them that in simple optics you can make an alternating focusing system by equal convex and concave lenses. The Alternating Gradient (AG) focusing idea was published near the end of 1952 by Courant, Livingston and Snyder. Much to their surprise, then it was found that the idea had actually been patented earlier by Nick Christofilos (see sidebar), who at the time was working in Greece as an elevator engineer.

This was in 1952, just at the time that European countries were joining together to form CERN in Geneva, Switzerland (see sidebar for CERN and Fig. 5.7). Their aim at first was to build a 10 GeV scaled-up version of the Cosmotron, and much of the design work had been done but, as if on cue, CERN visitors — Odd Dahl (see sidebar in Chapter I), Frank Goward and Rolf Wideroe (see sidebar in Chapter III) — arrived to see the Cosmotron. Hearing from the Cosmotron team of their new strong focusing idea, they immediately abandoned plans for a 10 GeV weak focusing machine in favor of a 25 GeV strong focusing proton synchrotron.

They were convinced that the reduction in the dimensions of the magnet gap and beam chamber would allow them to go to nearly three times the energy for the same price. Brookhaven had already planned such a machine as their next step — the Alternating Gradient Synchrotron or AGS (Fig. 5.8).

The first people to take up the idea of AG focusing were Bob Wilson and Boyce McDaniel (see sidebars for Wilson and McDaniel), who were then in the process of building one of the first electron synchrotrons at the Laboratory for Nuclear Physics, Cornell. They quickly modified the magnet design to try out the idea. Meanwhile the Brookhaven team wanted to satisfy themselves that the idea would work before embarking on the AGS, and built an electrostatic model. The leading figure in building the AGS was Kenneth Green (see sidebar for Green). Waiting for the results of this model delayed the AGS somewhat and enabled CERN to

Ernest D. Courant (1920–)

American physicist
Member of National Academy of Sciences, 1976
Enrico Fermi Award, 1986
American Physical Society R.R. Wilson Prize, 1987

Born in Göttingen, Germany, the son of a famous mathematician (the Courant Institute at New York University is named after him), Ernest Courant had a distinguished career in physics. He became an American citizen in 1940. He obtained a BA from Swarthmore College in 1940, then a PhD from Rochester in 1943. He spent 1943–1946 in Montreal at the National Research Council of Canada where he was concerned with the design of the first nuclear reactor built outside the United States. Then he worked under Hans Bethe at Cornell from 1946–1948, after which he went to Brookhaven National Laboratory where he remained until retiring in 1990. He was a professor of physics at the State University of New York, Stony Brook, from 1967–1985.

He has enjoyed time spent at many different universities, including Princeton (1950–1951), Yale (1961–1962) and Cambridge (1956–1957). He has also spent some time (1968–1969) at the National Accelerator Laboratory (now Fermilab). He played an important role as a consultant to that Laboratory when it was first set up.

Although Courant is best known for his invention of strong focusing, he has contributed to many other aspects of accelerator physics — for example, his work on the acceleration of polarized particles. He has had a vast role in developing the science of accelerators or, more precisely, the behavior of beams of particles in accelerators, storage rings and synchrotron radiation facilities. His sophisticated approach, bringing to bear the power of theoretical physics to what at first seems a hardware problem, has been an inspiration to all. It is no exaggeration to consider him the leader in the development of accelerator science.

Hartland S. Snyder (1913–1962)

American physicist
Co-discoverer of alternating gradient focusing
Co-discoverer of black holes

Snyder was born in Salt Lake City and graduated from the University of Utah. He did his graduate work at the University of California in Berkeley under J. Robert Oppenheimer, receiving his PhD in 1940. Together they studied what would be the final state of very heavy stars once the nuclear fuel is exhausted. They discovered that the stars would undergo gravitational collapse, and were thus the first to realize this possibility. Nowadays we believe that black holes are a very important part of the cosmos. Most astronomers believe that there is a black hole in the center of every galaxy (including our own); certainly there are black holes powering active galactic nuclei.

After leaving Berkeley, Snyder went to Northwestern University, and in 1947 he came, along with the famous physicist S.A. Goudsmit, then at the University of Michigan, to start the new Brookhaven National Laboratory. There, he contributed to the theory of cosmic rays, scattering theory, and quantum field theory. He also helped build and run-in the first high energy accelerator at Brookhaven; the Cosmotron. He was also active in the quest for controlled fusion energy, a quest that still goes on, with great progress having been made since that time.

He is best known for his co-discovery of strong focusing. He played a major role in the construction and study of the electron model of the Alternating Gradient Synchrotron (AGS), and in the design of the AGS. In 1960 he took a leave of absence from Brookhaven and worked on the Cambridge Electron Accelerator. He then went to Berkeley to work on what would later become the accelerator at Fermilab. He suffered many heart attacks during his last years, but used the time in the hospital to learn Russian (well enough to act as a translator to visiting Russian scientists). He finally succumbed to a heart attack while in Berkeley. Snyder was known for his boyish enthusiasm, good spirit, and cheerful manner.

Nicholas C. Christofilos (1916–1972)

American/Greek engineer
Inventor of strong focusing

Nick Christofilos was born of Greek parents in Boston, Massachusetts. The family returned to Athens when he was seven years old where he went on to graduate in electrical and mechanical engineering. As an elevator engineer he then set up his own firm but, as an intellectual recreation, found time to read textbooks on nuclear physics and high voltage phenomena and soon, even the *Physical Review*. It was during this time that he made two of his most important contributions: the strong focusing principle, and his scheme for plasma confinement.

He had developed an accelerator scheme not unlike the synchrotron as early as 1946, patenting it in both US and Greece. This invention was ignored since it offered no solution to the problem of phase stability. Undeterred, he went on to invent strong focusing, submitting patent applications in 1950 and sending copies to Berkeley. Apparently this was circulated, but its importance not appreciated.

Meanwhile strong focusing was reinvented by Courant, Livingston, and Snyder. It was in 1953, during a visit to the US and while reading in the Brooklyn Public Library, that he came upon their article in the *Physical Review*. He rushed to Brookhaven where he was shown round by John Blewett. John was convinced that Christofilos had discovered the principle independently and before the Cosmotron team; he was immediately offered a job working for Ken Green on the AGS.

Soon after he became interested in plasma confinement and his "Astron Proposal" in which cylindrical layers of high density electrons spiral around closed field lines and form a magnetic bottle. Somewhere in the wall of the bottle the field reverses direction and passes through zero. In this zero field layer circulating electrons are free to acquire considerable energy and, colliding with ions trapped in the bottle exciting and containing a very hot plasma.

The natural place for him to pursue this idea was Livermore but his Greek origins made it difficult at first for him to do so. When finally he joined Livermore he invented the induction linac to produce the Astron electron layer. He, in fact, built the whole Astron.

His imagination knew no bounds and in 1957, as part of his effort to improve National Defense, he proposed the creation of an artificial belt of electrons trapped in the earth's field lines — this was before the discovery of the natural Van Allen Belts. As an experiment he detonated a 2 kiloton nuclear bomb in the upper atmosphere to demonstrate, as he had predicted, a brief communications blackout. Another invention of his was the extremely low frequency communications system "Sanguine" used by the US Navy submarines.

Unfortunately, he died at an early age, at the very height of his powers.

Fig. 5.7 View of the CERN site in April 1957 during construction of the 26 GeV Proton Synchrotron (PS). The PS accelerated protons to 24 GeV in November 1959.

CERN

Without a conscious effort of imagination it is now difficult for us to appreciate just how divided the people of Europe were at the end of the Second World War. Although they had both the will and the need to stifle their hatred of each other, a decade of propaganda had taught a generation of English, German, French and Italians to be deeply suspicious of the motives of those outside their borders. Men of vision such as Jean Monnet, a founder of the European Iron and Steel Community — later to become the European Community — saw common enterprise as the balm to heal such political and psychological wounds of war.

There were those in the scientific community who seized upon the foundation of a European physics laboratory as just such an opportunity. Scientists had never found national frontiers of much importance, and the content of their deliberations, ultimately subject to the test of experiment, left little room for political opinion. Moreover, physics was finding its ambitious program of accelerators for higher and higher energy too expensive for any one country to afford alone.

As early as 1946 the French delegation to the United Nations asked for a study of United Nations Research Laboratories. This was referred to UNESCO where it was taken up at a meeting in Florence in 1950 by Nobel Laureate Isidor Rabi, supported by Auger from Switzerland and E. Amaldi from Italy. Rabi's enthusiasm was not entirely shared back home in Washington and so it was a collection of (almost) all European States that went ahead to sign the first CERN convention in 1952. Characteristically, the British had not joined; H.M. Government thought the machinery of scientific collaboration "unduly complex." It took more than a year and the efforts of Lord Cherwell and Sir Ben Lockspeiser to change the Government's mind. Lord Cherwell's intervention is somewhat ironic since only a few years earlier, as Churchill's advisor, he has been blamed for encouraging him to bomb civilian targets such as Dresden. Perhaps CERN's purpose of reconciliation had already started to take effect!

In 1952 Geneva, Switzerland was chosen as the site of the new Laboratory over alternative proposals in France, Denmark and the Netherlands. Switzerland: a neutral nation with its tradition for hosting international movements is located at the geographical centre of gravity of Europe. Other contending nations had their counter-arguments but failed to win the day.

The first Director, Cornelius Bakker of the Netherlands, was appointed in 1954 in time for the first CERN Council meeting, and work started almost immediately on the construction of the 25 GeV proton synchrotron. A synchro-cyclotron had also been designed as a tooth-cutting exercise and was completed before the PS.

The 25 GeV PS was followed by the Intersecting Storage Rings, a highly successful project that had to navigate uncharted waters, overcoming numerous beam instabilities that threatened its performance. The Super Proton Synchrotron (SPS) followed: a rival to Fermilab's kilometer radius main ring. Rather than compete with Fermilab's plans to augment the 450 GeV energy of its machine by installing a ring of superconducting magnets, CERN boldly took on the conversion of the SPS into a collider for counter-rotating beams of protons and antiprotons. Accumulating a sufficient number of antiprotons and cooling them to the density needed for high luminosity stretched the imagination and skills of CERN's scientists and engineers to the ultimate, but won them their first great achievement, providing the facilities that led to a Nobel Prize awarded to Carlo Rubbia and Simon van der Meer, for the discovery of the W and Z bosons.

CERN went on to construct LEP, an electron-positron collider roughly four times the circumference of the SPS, and marked the turn of the millennium by using LEP's tunnel to house the largest proton-proton collider ever — the LHC. All these great engines of discovery are described elsewhere in this book.

CERN has become more and more a world laboratory, particularly since the discovery of the W and Z in the early 1980s. It has grown to be host to thousands of visiting scientists, formed into huge collaborative experiments which focused on the large detectors of the SPS proton-antiproton collider. These were followed by those of LEP and finally those of the LHC — about to be completed. The aim of bringing together Europeans was an effortless success from the very start and has now been extended. A succession of countries aspiring to be members of the European Community have viewed joining CERN as a useful step towards their candidature.

American scientists whose experiments at the great laboratories in the USA used to dominate progress in the field now seem to turn naturally to CERN for the facilities they need. CERN has vigorous collaboration agreements also with Russia, China, India, some South American countries, and other countries not strictly within Europe's boundaries.

---

overtake them. In Fig. 5.9 we see the first CERN magnet unit being installed.

V.8 Brookhaven's AGS and CERN's PS

By 1959 CERN's PS was ready for testing, somewhat ahead of Brookhaven's AGS, but there had always been some doubts about what would happen to the beam at "transition," a sort of watershed in the acceleration process when an effect that kept the beam in step with the accelerating field changed from stable to unstable. Indeed the CERN PS faltered at transition until a bright young radio engineer, Wolfgang Schnell (see sidebar in

Chapter X), produced a circuit he had built in a Nescafe tin to change the phase of the accelerating wave at the moment of transition. Schnell, his box, and a few coaxial connectors brought the beam through transition to full energy.

John Adams (see sidebar for Adams), who had led the CERN team with a nice mixture of courage and caution, was able to announce their success on 24 November 1959. Key members of the team were Mervin Hine and Hugh Hereward (see sidebars for Hine and Hereward). Colleagues at Dubna in the USSR had sent CERN a bottle of vodka with which to celebrate. The bottle was dutifully returned empty, with a message of thanks.

Fig. 5.8 An artist's impression of the AGS before its construction in the 1950s.

Fig. 5.9 The first Proton Synchrotron (PS) magnet unit, with members of the Magnet Group. This was one of the very first machines at CERN, built in the 1950s, and still the central element in the very extensive CERN complex of accelerators.

Robert Rathbun Wilson (1914–2000)

American physicist
Constructed a series of synchrotrons at Cornell
Was the first director of, and built, Fermilab
Pioneered the use of accelerators in medicine
National Medal of Science, Wright Prize, the Enrico Fermi Award
President of the American Physical Society, 1985
Member of the American Academy of Arts and Science and the
  National Academy of Sciences, 1957

He was, for many of us, the most important person in accelerator development after Ernest Lawrence. His contributions to accelerator science are beyond belief. Born in Wyoming of rancher parents, and with early childhood spent in "frontier" America (which shaped his personality), he suffered from age 8 on as his parents separated and he attended many different elementary schools in Illinois. Nevertheless, even at that time, he indicated an interest in things mechanical. He attended the University of California where he graduated in 1936 and received a doctoral degree in 1940. He studied under Lawrence and, even as a graduate student, made important contributions to accelerator technology. He then went to Princeton and was involved in this University's short-lived project for eletromagnetic separation of uranium. After that he went to Los Alamos and was in charge of the cyclotron laboratory and, subsequently, made chief of the Physics Research Division.

After the war Wilson played an important role in seeing that atomic energy was under civilian control. He was instrumental in initiating the Federation of American Scientists and continued, throughout his life, to be involved with arms control matters.

In 1946 Wilson went to Harvard, where he stayed only two years, but built a cyclotron and first suggested the use of fast protons in cancer therapy. That idea was of seminal importance; nowadays there are an ever growing number of proton therapy machines. The cyclotron he built continued to be used to treat cancer patients for a number of decades.

In 1947 he went to Cornell, where he built the Laboratory of Nuclear Studies and a series of four electron synchrotrons, ranging from 300 MeV to 12 GeV. He was the first to experimentally check out the revolutionary concept of strong focusing.

In 1967 he became director of the newly formed Fermilab (then called the National Accelerator Laboratory). There, he built the accelerator at 30% lower cost than the first proposal. At the same time his flair for things artistic was evident in the structures and even the design of the electric power lines. He, himself, was quite an accomplished sculptor. He was creative and imaginative in ways that are hard to believe. For example, a section of Fermilab was made to revert to original prairie grass that soon attracted rare species of birds. A herd of buffalo was also incorporated.

Perhaps Bob Wilson is best remembered for his wonderful response, in 1969, to questions by Senator Pastore:

Senator: "Is there anything connected with the hopes of this accelerator that in any way involves the security of the country?"

Wilson: "No sir. I don't believe so."

Senator: "Nothing at all?"

Wilson: "Nothing at all"

Senator: "It has no value in that respect?"

Wilson: "It has only to do with the respect with which we regard one another, the dignity of man, our love of culture. It has to do with; are we good painters, good sculptors, great poets? I mean all the things we really venerate in our country and are patriotic about. It has nothing to do directly with defending our country except to make it worth defending."

Boyce D. McDaniel (1917–2002)

American physicist
Director of the Laboratory of Nuclear Studies, Cornell
Trustee of Associated Universities and the Universities Research
  Association
Member of the National Academy of Sciences

Born and schooled in Brevard, North Carolina, he earned his bachelor's degree from Ohio Wesleyan University in 1938 and his master's degree from the Case School of Applied Science (now the Case Western Reserve University) in 1940. McDaniel finished his doctoral thesis at Cornell in 1943, under the direction of Robert Bacher, who then brought him to Los Alamos. At Los Alamos, during the war, he played a key role in Wilson's cyclotron research team. McDaniel was the last man to check, and even touch, the test atomic bomb, at the Trinity Site, before it was detonated at 5:29:45 a.m. Mountain War Time, July 16, 1945.

In 1946, McDaniel joined the Cornell faculty as an assistant professor and became a full professor in 1955. He was a leader in establishing the Laboratory of Nuclear Studies and had a leading role in designing and building the 300 MeV electron synchrotron, one of the first such accelerators in the world. He and Robert Wilson, who was McDaniel's predecessor as director of LNS, built three more electron synchrotrons of successively higher energies, each of which enabled physicists to study phenomena in a new energy range. McDaniel became director of LNS in 1967 and remained in that position until he retired from the Cornell faculty in 1985. He pioneered the technique of tagged gamma rays and performed important measurements with each of these accelerators, including a long series of work in K-meson and lambda-meson photo-production and measurements of the neutron electromagnetic form factors.

In 1972, McDaniel took a year's leave from Cornell to become acting head of the accelerator section at Fermilab in Batavia, Ill. Though the Fermilab accelerator had operated at near-design energy, component failure was frequent and operation intermittent. McDaniel had the accelerator working reliably by the end of the year.

George Kenneth Green (1911–1977)

American physicist
Constructed the AGS at the Brookhaven National Laboratory

After the discovery of strong focusing at the Brookhaven Research Laboratory in the United States, Ken Green was put in charge of building the first alternating gradient proton synchrotron, the AGS, and it was he who proudly announced its first operation in July 1960. Ken had quite a different personality from Adams, who was then his counterpart in the rival CERN project. It would be hard to improve upon the description of Ken by Mel Month, who worked closely with him: "Ken had that rare combination of dedication to detail and an expansive personality that virtually oozed persuasiveness. He was at once an eager beaver and a wheeler-dealer. He could talk your head off and, in the end, if it wasn't his argument that got to you, his charming way of holding his lighted cigarette would." Ken was aware of accelerator matters throughout the country and very willing to offer his opinions. For example, he heard the Argonne people promise high intensity from the ZGS. (Higher than from any AG machine, such as the AGS, because of the large aperture of the ZGS.) He bet that the AGS would reach a higher intensity than the ZGS and he won the bet.

The Brookhaven National Laboratory in the 1960s suffered from the dichotomy and conflicting interests of the accelerator designers, headed by Ken, and the research physicists under the direction of Maurice Goldhaber and later George Vineyard. The AGS, thanks to Ken and his team and its physics program, was at first unrivalled in its success, but Brookhaven's plans to build the next big machine were frustrated by the politics of the rivalry of the fund-giving agencies and of other aspiring laboratories. Ken arrived at decisions by following rational rather than political paths.

When the time came in 1963 for a national panel under Ramsey to decide what machine to build next, it became clear that this was to be a 200 (later 400) GeV proton machine. There had been an agreement between Berkeley and Brookhaven that if the AGS went ahead, the next big machine would be at Berkeley on the West Coast. As it turned out the project was finally given to Bob Wilson to build in Illinois, at Fermilab; but either way Ken Green had to look elsewhere for something to motivate the aspirations of his team. The plan he worked out with Berkeley was that Brookhaven would occupy itself with the R&D for the collider to follow the 200 GeV fixed target machine.

Superconducting magnet technology seemed to be the way of the future and Ken was keen to follow it. However, his physics colleagues at BNL, who seemed happy to milk the AGS for all it was worth and let the 200 GeV machine go elsewhere, were not keen on letting Ken have funds to design a collider. They considered this to be inferior to the fixed target alternative and saw superconducting magnets as a long way off. But when the time came in 1970 for John Blewett, Ken's second-in-command, to start superconducting R&D for ISABELLE, Ken had already fallen victim to political infighting.

He was not allowed to continue beyond 1969 as head of the AGS, a cruel blow; and, as many have done before him, he tried to make the best of a bad world. The couple of years before his death in 1977 found him designing a solar steam plant to supply process steam to the output of the Brookhaven Laboratory's oil fired central heating system to save the 7 million gallons of oil that it burned annually. At this time he was also preparing the design of the National Synchrotron Light Source Project (NSLS) for Brookhaven. Happily in this context he is remembered along with Renate Chasman for the invention of the ingenious "Chasman-Green" lattice cell — an arrangement of lenses and bending magnets that has been used by most designers of synchrotron light sources as the basic unit of their ring design.

Unhappily he died in 1977, just before funding for the NSLS was granted to BNL.

The AGS followed rather soon afterwards and was for a short time better prepared to launch into a rich programme of experiments than CERN.

V.9 Fermilab and SPS

After the AGS and CERN's PS there had to be a significant leap in beam energy. A factor of 10 would raise the "fixed target" energy by a factor of 3. Somewhere between 200 and 300 GeV seemed to be a reasonable aim, provided economies in the construction could be implemented. The mastermind behind the first of these second-generation proton synchrotrons was Robert (Bob) Wilson, who had worked with Lawrence in the cyclotron era and had built a number of successful electron synchrotrons at Cornell. He had no hesitation in adopting the AG principle for these machines and he was to add his own particular flavour to the construction of his new 200 GeV machine at the Fermi National Accelerator Laboratory (FNAL) near Chicago (see box

Sir John Adams (1920–1984)

English engineer and physicist
Constructed the CERN Proton Synchrotron and Super Proton
  Synchrotron
Twice served as Director General of CERN
Fellow of the Royal Society

John Adams was the "father" of the giant particle accelerators that have made CERN a leader in the field of high-energy physics. He was an extraordinary accelerator designer, engineer, scientist and administrator (without any formal qualifications).

His style was that of a gentleman. He chaired meetings by listening quietly to all the arguments while excavating the hidden passageways of his pipe. Then would follow a period of solitary cogitation as, with meticulous handwriting, he would analyze every problem and plan his response in the pages of his daybook. Others would then have to be persuaded to accept the plan. With patient but authoritative words he would convince them of the inevitability of his logic until later they themselves might propose the solution as if it were their own. Honesty and transparency were his watchwords.

He worked during World War II in the Radar Laboratories of the UK Ministry of Aircraft Production and then at the Atomic Energy Research establishment at Harwell on the design and construction of a 180 MeV synchrocyclotron. He came to CERN in September 1953 and was appointed director of the PS division in 1954 at the age of 34, becoming the leader for the world's first big international particle accelerator project.

From 1961–66 Adams worked as director of the Culham Fusion Laboratory but in 1968 he came back to CERN — his mission — to revive the European "300 GeV Project." With patient diplomacy he was able to persuade CERN's Member States, many of whom wanted the project on their own soil, to place the machine at CERN. Among those to be persuaded was the UK's Margaret Thatcher. The iron lady was so impressed by his arguments that she telephoned him in person to let him know when the UK reversed their decision to stay out.

His skill at forming and directing a team, which was a judicious mix of experience and youthful enthusiasm for innovation, resulted in a highly successful machine. The SPS was completed on time and within budget — adding to his already great reputation at CERN.

Once the SPS was finished he went on to again become Director General of CERN and laid the groundwork for the Proton Antiproton Collider, which led to the discovery of the intermediate vector bosons.

Mervyn Hine (1920–2004)

British physicist
Designed and built the CERN's PS with Adams
Member of CERN's Directorate for many years

Mervyn Hine was born in Hertfordshire, England. His father was a bacteriologist who did research at St. Bartholomew's Hospital, London. He went to Gresham's School in Norfolk where his short-sightedness and a severe stammer set him apart from his fellow pupils and made him take refuge in mathematics and science. He went on to King's College, Cambridge, from which he graduated at age 19. That was during World War II, but rather than being called up for military service he was sent to work at the Telecommunications Research Establishment in Malvern.

Hine had a particularly incisive mind. With a first class degree from King's College, Cambridge he worked together on radar research in Malvern with John Adams who was to become his colleague at Harwell and later at CERN. The two complemented each other. Adams, who had no formal training was skilled at the practical side of building equipment and was good at getting his own way with people. Mervyn's strength was his mathematics. As he himself said, he also had the courage "to say that which no one else was prepared to say" — and perhaps that which Adams would rather keep to himself.

After the war they worked together at the Harwell cyclotron, where they published a seminal paper on the management of resonances that pointed the way forward in the design of big machines. The "terrible twins," as they came to be known when they moved to Geneva in the early days of CERN's PS, were an outstanding pair of leaders for the first generation of CERN machine physicists and engineers.

After the PS was finished Adams left CERN but Mervyn Hine remained and gravitated, like many before him, into the higher management of the organization. He had a way with accounts, inventing procedures to set out the financial plans for the CERN Council in a fashion that was transparent, logical and convinced the delegates that they would be taking no risks with their governments' science budgets if they approved the projects that CERN proposed. At times there might be three such projects overlapping in a "four-year plan" all involving an increased expenditure. His approach relied on applying the rigid rules of calculus to the budget: using percentage increase of prosperity to discount future growth of the budget, thus leaving room for new capital projects.

Returning to CERN in 1970, Adams, the driving force behind one of these projects, the SPS, relied heavily on Hine's financial wizardry with the Council — although he bemoaned the loss of Hine's intellect to the physics community. He once said, "You know Mervyn has an excellent brain and I don't know why he wastes his time in finance."

Once the kind of financial approach he had pioneered became commonplace, Hine returned to the technical domain. From 1964-71 he supervised CERN's computer development by arranging for CERN to receive the first CDC 6600, after endless discussions with Seymour Cray, the American designer of the machine. He was also involved in numerous high-level technical negotiations with multinational suppliers, always arguing that it was a major development opportunity for them to supply CERN — an opportunity for which they should pay!

Mervyn loved *l'esprit français* — he regularly read *Le Monde* and *Le Canard Enchaîné* and knew every bit of French official politics and gossip. He had an intense love for music and sang in classical choirs, first at school and later with Hildred Blewett, Ted Wilson and many other colleagues from the accelerator world, during his days at CERN. His sparkling and witty intellect remained with him to the end when sadly he died at his home in Founex, Switzerland.

Hugh Hereward (1920–)

English accelerator physicist
Leader of Machine Studies for CERN's PS and ISR

Hugh Hereward was born in London. His father had been an ocean-going captain of merchant vessels until he left the sea to become manager and owner of a precision engineering firm in Birmingham founded by Hereward's grandfather in the mid 1800s. Hugh Hereward attended the King Edward's High School in Birmingham, where a physics teacher who believed in Hereward's abilities encouraged him to specialize in the subject. He won a scholarship to St. Johns College, Cambridge, where Cockcroft was at the time.

Like Mervyn Hine he took a shortened degree course and in 1942 chose to continue research after graduation rather than join the army. He was one of several Cambridge physicists who contributed to atomic energy development in the Second World War. In 1942 he was in Halban's group at Cambridge, moving to Montreal in 1945. Later he was to become one of the founding fathers of CERN.

On first meeting he seemed a rather reserved and eccentric person. Although married, he had the uncared for appearance of a bachelor who had still not become used to choosing clothes for himself since he shed his school uniform. His well-worn flannel trousers were supported by a broad leather belt whose tarnished buckle was fashioned into the insignia of the Boy Scouts, complete with the motto "Be Prepared." We dared not ask if this was a reminder to himself or an exhortation to others.

After joining CERN in 1953 he led the group which built the first 50 MeV linear accelerator used to inject protons into the PS.

Hugh — or Hereward, as he was invariably referred to in those more formal days — was a great intellect whose chosen role was to study and analyze whatever knotty problems in accelerator theory were preventing a full understanding of how the PS, and afterwards the ISR, could be operated more effectively.

He saw things very clearly and produced a number of classic memos on matching and mismatching beams, instabilities, stochastic cooling and Landau damping. These were often circulated in handwritten form but were so clear that his colleagues, when asked about these subjects forty years on, will still search their filing cabinets for "something that Hereward wrote on this." He invented the first device to measure the current in a beam using the transformer principle as well as a method of stabilizing the oscillations of a bunch of particles due to a timing error as they pass through the accelerating cavities of a synchrotron; both these inventions bear his name.

Colleagues would often consult him — the experience was reminiscent of a trip to Delphi. The question would be posed and a silence would follow — often long enough to induce unease in the questioner, who might be tempted to rephrase the question. This was unwise, for the clock would be restarted and another long pause would ensue: Hugh would seem disturbed that his thought process had been interrupted. Usually the ensuing verdict would be delivered with gravity and precision but, failing that, the question would rebound in quite a different formulation, requiring a day or two's further research on the part of the questioner.

Hugh Hereward's heyday was when he led the Machine Studies Team that developed the performance of the PS in the 1960s — he became leader by common acclaim and went on to fill the same role in the ISR. Although he made many contributions to this new storage ring and to the theory of beam cooling, he seemed less at home in the ISR and it was not long before he retired from CERN, to live quietly in a Dorsetshire village, while occasionally accepting consulting engagements for other laboratories in Canada and Europe.

on Fermilab and Figs. 5.10 and 5.11). As director and project leader of this new laboratory he brought to the enterprise a flavour of economy and innovation he had inherited from his mentor Lawrence. By separating the functions of focusing and bending, which had been combined in the magnets of the AGS and PS, and using pure quadrupole and dipole units, he found he could squeeze more bending power into each kilometre of the ring.

Wilson's team were encouraged to use lateral thinking to emulate their leader's unconventional approach. Key members of Wilson's team were Boyce McDaniel, Paul Reardon and Roy Billinge. (See sidebars for Reardon and Billinge.) One such innovation, due to Tom Collins (see sidebar for Collins), was to use a "telephoto" system of magnetic quadrupoles to shoot the beam across a 60-meter-long straight section, catching it at the other end with the mirror image of this optical system. These modifications were made at six points in the ring to house crucial components for injection, acceleration and extraction, which needed

Fermi National Accelerator Laboratory (FNAL) near Chicago

In 1967 a large tract of empty farmland on the outskirts of Chicago was chosen as the site for a new accelerator. It was to be a synchrotron, twenty times the energy of AGS at Brookhaven and CERN's 25 GeV Proton Synchrotron. It had to be a factor of twenty because in those days of fixed target accelerators this brought only a factor of the square root of twenty — just over four — times the available energy for creating new particles. This seemed to be the minimum worth aiming for. With clever design the ring would only be ten times bigger but one could no longer hope to find a large enough site on a university campus or within an urban environment.

Much of the design work for the new accelerator had been carried out on the West Coast at Berkeley, California. It was therefore somewhat of a disappointment for those who had worked on the design to find they might have to move to a less temperate clime. Chicago, the most American of cities, is by no means philistine, but wives and families of potential recruits found the climate very much a second best to California.

The other surprise was the choice of Robert Rathbun Wilson (Bob Wilson) as the Director. Although Bob had been one of Ernest Lawrence's brightest protégés, and had built a series of electron machines at Cornell, where he had been the first to try out the revolutionary alternating gradient idea, he was somewhat in the shadow of the Berkeley and Brookhaven experts who already had proton machines of some stature in their logbooks. The reader will find a pen-portrait of Bob Wilson on page 66 and may judge for themselves the wisdom of the choice. However, anyone who witnessed those early years of Fermilab (as it later came to be known) cannot fail to applaud the effect of his leadership.

Fermilab's 6,800-acre site was originally home to farmland, and to the village of prefabricated homes called Weston. Bob converted them to offices and workshops but kept some of the original barns as a club house and social centre where his hard working team might relax and regroup for the week to come. Soon the other antique wooden farmhouses on the site were dragged bodily to form a small hamlet around the old village — and eventually become a community for the international world of physics. Among Wilson's early imprints on the lab was the establishment of a herd of American bison, symbolizing Fermilab's presence on the frontiers of high-energy physics, and the connection to its prairie origins. He recruited a Native American to look after the bison herd, and when they had to be culled the buffalo would be roasted for a lab party. The herd stands today, and new calves are born every spring.

Bob's first technical task was to define the project which was to put the US in the forefront of physics for a couple of decades — and as we write, until CERN in Europe fills the void after LEP with the LHC, Fermilab is the only laboratory able to make a major discovery at the high-energy frontier.

A machine of such dimensions could hardly be built to the exacting standard of its predecessors and so every possible economy of size and "comfort factor" had to be pared down to the minimum for the first-high energy ring (400 GeV). This was to become the style of the lab. If it worked the first time it was over-engineered! This certainly hurried the project along — even counting the time to fix things afterwards. The other attribute of Bob's style was aesthetic excellence. As a sculptor he took a hand, and had particular interest in the design of many structures and, most specifically, of the imposing "High Rise" building. Fermilab today boasts fine architecture punctuated by his striking sculptures and bold eye-catching vistas.

Cost-cutting has its price, however, and the first Main Ring was therefore not without initial disappointments. Anyone responsible for a shift in the control room in the early days might count themselves lucky to survive the night without a magnet to be replaced or a fire to be extinguished in a power supply. But those were the early days: before long — and still in a timeframe that others might have deemed impossible — the lab lived up to its promise.

There followed a few lean years as Bob Wilson — anxious to raise funds for a more powerful superconducting ring to reach 1 TeV (1000 GeV) — went too far in putting his own Directorship on the line. Enemies among the trustees who would have rather seen the money go to Brookhaven called his bluff — sacked him and so lost the most charismatic and inventive accelerator builder of all time.

Fermilab survived, however, and under the Direction of Leon Lederman went forward to build the Tevatron — the first large superconducting machine — and (rather belatedly) to use it like CERN's SPS as a proton-antiproton collider. The Tevatron, four miles in circumference and originally named the Energy Doubler when it began operation in 1983, is the world's highest-energy particle accelerator, and will remain so (until the LHC is turned on in 2008). Its 1,000 superconducting magnets are cooled by liquid helium to −268°C (−450°F). Its low-temperature cooling system was the largest ever built when it was placed in operation.

Two major components of the Standard Model of Fundamental Particles and Forces were discovered at Fermilab: the bottom quark (May–June 1977) and the top quark (February 1995). In July 2000, Fermilab experimenters announced the first direct observation of the tau neutrino, the last fundamental particle to be observed, filling the final slot in the Standard Model.

To Fermilab's chagrin, it was never given the chance to provide the environment (and its existing machines) for the SSC. However, in all fairness it must be said that the decision to put the SSC in Texas was very much in the "green-field" spirit that had engendered Fermilab in the first place, but the non-choice of Fermilab was one of the important cards stacked against the ill-fated SSC.

In recent years Fermilab has added the two-mile Main Injector accelerator to increase the number of proton-antiproton collisions in the Tevatron, greatly enhancing the chances for important discoveries. It has also embarked upon an extensive neutrino program with NUMI and MINOS and a third possibility on the horizon. It remains the focus of the aspiration of many young and pioneering minds still hoping to use the Tevatron to find the Higgs Boson before LHC is finished, seeking to determine whether or not there is a sterile neutrino, and studying properties of oscillating neutrinos.

Fig. 5.10 A striking view of the Fermilab "High-Rise" building — named "Wilson Hall" in tribute to Bob Wilson's creative genius.

Fig. 5.11 The Fermilab machines as seen from the air. Behind the new main injector ring in the foreground we see the ring tunnel that was built for the first large accelerator: a 400 GeV conventional magnet synchrotron.

Paul Reardon (1930–)

American physicist
Pioneer of superconducting magnets for synchrotrons and tokamaks

Paul Reardon graduated from college in Boston in 1952 and within the short space of a week took a commission in the US Army Reserve, married his lifelong partner Pauline, and started work as a junior physicist at the Johns Manville research center. The Korean War was raging at that time and Paul's new career was immediately put on hold while he spent 2 years as a gunnery officer in Japan and Korea, earning the Bronze Star for his efforts.

Back at the research center he applied basic science to the problem of squealing brakes on subway trains and trucks. The new material he found for the brake shoes has made the world a quieter place for more than half a century.

In 1958 Reardon moved to Princeton where he learned physics and enjoyed his many personal interactions. Perhaps due to this experience, but more probably as a result of raising a happy family of seven children or, perhaps, as a result of early training at the hands of the Jesuits, Paul's management style is exceptionally humane.

At Princeton he became head of the Magnet group at the Princeton-Penn accelerators. In 1964 he moved to the Division of Research of the Atomic Energy Commission (USAEC). From 1966 to 1969 he was the Project Manager for the Bates 400 MeV Electron Linear accelerator at MIT. In 1968 he was recruited by Bob Wilson to come to Fermilab, where he was head of the Booster Synchrotron Section, Director of Business Administration and then Head of both the Accelerator Division and the Superconducting Energy Doubler Accelerator Section.

After Fermilab Paul moved from accelerators to devices for Controlled Nuclear Fusion. In 1975 he became Project Manager of the Tokamak Fusion Test Reactor and, in 1982, became Associate Director of the Princeton Plasma Physics Laboratory. In 1983 he moved back to accelerators — to Brookhaven — where he headed the Relativistic Heavy Ion Collider (RHIC) from 1984 to 1988 before becoming Project Manager for the ill-fated Superconducting Super Collider (SSC). From the 1990s on he was interested in the development, design and construction of small proton accelerators for cancer therapy.

Throughout this varied career he has always been a highly visible (and entertaining) participant in the US and international high-energy physics and fusion energy programs. He has been in a large number of review and advisory panels to which he brought his abundant common sense and excellent social skills — always rising above the squabbles that sometimes plague discussions between rival experts.

Roy Billinge (1937–1994)

English accelerator physicist
Built the SPS magnet and the CERN Antiproton Accumulator

Roy Billinge was unusual among accelerator physicists in having worked for two giants of the accelerator world, Bob Wilson and John Adams, in the full flood tide of their major projects. Both men, despite their very different styles, held him in the highest regard.

Born in Buxton, England, Billinge studied physics at King's College, London, and joined the UK Rutherford Laboratory in 1959 where he immediately became fascinated with building accelerators, playing an important role in bringing the 7 GeV NIMROD proton synchrotron into operation. He went to CERN in 1966 for a year, seconded from the Rutherford Laboratory to the "300 GeV project," which went on to become the SPS proton accelerator. In 1967 he left to spend four years at the new Fermilab accelerator then being constructed under Bob Wilson.

Once his task, the design and completion of the 8 GeV Booster, was over at Fermilab, he was asked by John Adams to lead the magnet construction for CERN's new SPS. He was, at 33, the youngest Group Leader in a young team.

Billinge brought back to the Old World the spirit of adventure of the New, as well as a broad experience of project management on both sides of the Atlantic. He had a style of building a team into a family in which each member felt of equal importance. This became invaluable when the SPS was completed and he, together with Simon van der Meer, drew together diverse talents from all over CERN for the challenge of building the Antiproton Accumulator. Billinge, who was always ready with a suggestion to simplify and resolve a setback, pushed the AA from approval in 1978 through to commissioning in 1980. Then for nine years (1982–90) he was leader of the PS division.

In his remarkable mastery of humane and effective leadership, the main ingredient was his concern to understand and satisfy the needs and ambitions of the team of scientists and engineers under his charge. His wide professional experience and his gift for finding clear solutions that were mathematically correct, yet practical and sound in engineering terms, earned everyone's respect.

In the last years before his untimely death he helped in bringing a larger community of nations, including the USA, behind the construction of CERN's LHC.

Thomas Collins (1921–1996)

Canadian physicist
Member of the High Energy Advisory Panel to the DOE and the NSF
American Physical Society R.R. Wilson Prize, 1994

Collins was educated at the University of British Columbia, receiving his BS in 1941, his MS in 1942 and his PhD in 1950.

He was one of the main architects of the Cambridge Electron Accelerator (CEA). The CEA was led by Stanley Livingston, but his first hires were two great accelerator physicists, Ken Robinson and Thomas Collins. The two of them designed and built the machine, with Collins responsible for most of the major components, including the magnets, vacuum, and the general layout. His broad understanding of accelerators made him a "folk hero" to others (an air he enjoyed and cultivated). During the time he was at CEA, Collins developed the concept of long straight sections (and even patented the concept). Since then, many kinds of straight sections have been developed for injection and extraction into accelerators, for "wiggler inserts" in synchrotron radiation facilities (for making intense x-ray beams), and for special low-beta regions of colliders. It was, however, Collins who first realized the possibility.

Collins was one of the first who appreciated the importance of computers for accelerator physics, and used them everyday even in the 60s, when computers were very primitive.

After leaving the Cambridge Electron Accelerator, he joined the Fermi National Accelerator Laboratory in 1968 and remained there until his retirement in 1988. Tom Collins contributed in many areas during Fermilab's evolution from the Main Ring and its experimental areas, to the Doubler and the antiproton ("pbar") source, as well as the Collider Detector at Fermilab (one of the two major detectors on the Tevatron), and also the SSC. He also served many internal committees and on occasion served as acting Director for R.R. Wilson. His expertise was recognized within Fermilab (he was promoted to many important positions such as Associate Director for Accelerator Planning Advice) and well-known to the community of accelerator physicists, but, unfortunately, never rose beyond that.

He was a bright, creative person, but a loner. He had difficulty interacting with others; he tended not to ask others for advice, and he was not a good listener. On the other hand, he was clever and enjoyed running rings around those more formally elite. His home, during his CEA days (a great old house that required lots of repairs), was in a rather poor neighbourhood (compared to where others at Harvard and MIT lived). He was very involved in local politics and worked on local issues such as improving the public schools. He had many talents besides accelerator physics even having done the lighting for a Bolshoi Ballet performance in 1975. He was a hiker and a skier.

Fig. 5.12 Inside the tunnel of the SPS, built at CERN in the 1970s. The bending dipole magnets are painted red and the focusing quadrupoles are blue. On the yellow support frame we see some smaller magnets for tuning the machine.

plenty of space to do their job. With this and other bold economies such as reducing as many as possible of the gauges, taps and switches that threatened to adorn each of the ring's 1000 magnets, not to mention applying the production line techniques of Henry Ford to the construction, Wilson was able to double the energy target to 400 GeV and propose its completion in a mere 5 years. The first bold innovators in such ventures are perilously exposed, but he kept his promise to complete the machine in only 5 years and thereby gain a march on the rival SPS (see Fig. 5.12) at CERN, which had become stuck in a political quagmire and which started construction only as Wilson's machine began to run. Later the addition of a superconducting ring, the Tevatron, was to complete the world's first superconducting hadron collider at Fermilab.

Superconducting Magnets

When particles are bent in a circular path around an accelerator or collider, the radius of the path, and hence the sheer scale of the device, increases in direct proportion to the energy of the particle but in inverse proportion to the strength of the magnetic field that causes the bending. Clearly the magnetic field should be as strong as is practically possible, but for ordinary magnets, excited by copper coils, and whose field is produced between parallel iron pole-pieces, the iron yoke that returns the magnetic flux saturates at a field of about 2 T (tesla). To keep increasing the field after saturation, very much more current would be needed, because the iron yoke, previously so willing to accept the magnetic flux, now impedes the passage of the field as if it is made of air instead of iron. Below saturation the coil has only the small air gap between the poles to contend with but at higher field levels it must overcome the "reluctance," as it is called, of the whole flux path.

To force the field around this much larger "circuit" of air, the currents in the coils must be much higher and the power dissipated in the copper unrealistically so. Even below saturation an accelerator magnet ring can dissipate 50 megawatts of power.

By making the coils out of certain materials and cooling them to only a few degrees above absolute zero (273 degrees below freezing or 0 K (degrees Kelvin)) the motion of the electrons that carry the current through the coils becomes suddenly more ordered and the scattering of these electrons, which is the reason why power is dissipated (so called "ohmic" heating), ceases altogether. The electrons in a superconductor are subject to a great quantum mechanical unifying force, so if any electron hits an atom and tries to be scattered, the others hold it back. This effect is known as superconductivity and a coil wound from a superconducting alloy (niobium-titanium) and cooled to a few degrees Kelvin will drive a high magnetic field with little or no energy dissipation.

The only difficulty is that the magnetic field itself can destroy the superconducting properties. But with a suitable alloy and a low enough temperature, magnets of 4 T, twice that of conventional "room temperature" magnets, may be designed. The magnets of the latest superconducting accelerator, LHC, reach 8 T and for such a high field the superconductor had to be cooler (1.8 K instead of a more traditional 4 K). At this temperature the liquid helium that cools the coils has interesting new properties. It has almost perfect heat conduction — a good thing — but is super-fluid, so care must be taken in pumping it around and in the liquefaction process.

The shape of the field and its purity, so important for an accelerator, is no longer fixed by the precise profile of the iron pole pieces, but by the configuration of the coil, whose strands must be wound with the precision of a watchmaker.

Each coil is built up of layers of conductor around a circular form in a geometry that approximates to the "cosine theta" shape ideally needed for a uniform field. The layer of coils is thick at the top and bottom and tapers to nothing in the mid plane of the magnet. The main engineering challenge of making these magnets is to clamp the coils into this very precise shape, since the mispositioning of a conductor by a fraction of a millimeter will limit the life of the circulating beam. It is also true that even the slightest motion of a superconducting strand produces enough local heat to cause the magnet to "quench" (lose its superconducting properties).

The forces tending to blow the coils apart are huge and are contained by interlocking stainless steel collars around each coil. The whole assembly must be pre-stressed at room temperature so that it does not fall apart as it contracts.

Today's superconducting magnets are designed such that the field at which they quench is as close as possible to the field at which a short sample of superconductor would. This field for niobium-titanium is typically 6 T at 4 K and 8 T at 1.8 K. Niobium-tin has an even higher quench field but is brittle and must be formed in situ.

The critical temperature (T_c) of the superconductor is that at which it will quench no matter how low the field. High-T_c superconductors have received a great deal of attention since they were discovered by J. Georg Bednorz and K. Alexander Muller (Nobel Prize, 1987). So far, despite considerable effort, designers have not produced an accelerator magnet that is competitive with a low-T_c superconducting magnet (running at about 4 K). However, there is already a very important application of high-T_c superconductors to accelerators; as a feed-through to bring electrical power into cryogenic vessel. Ceramic high-T_c superconductors are perfect conductors of electricity but very poor conductors of heat — just perfect for this application.

Accelerator builders have pioneered the new and exciting technology of superconductors on an industrial scale even ahead of its use for underground power transmission and for frictionless levitated railroads.

V.10 Superconducting Magnets

Over the decades, as synchrotrons have become larger, physicists have continued to ask for beams of higher and higher energy. Their size is roughly proportional to the energy (strictly the momentum) of the particles they accelerate and is limited by the maximum field (about 2 T) that an electromagnet can produce before the iron parts saturate. An alternative is to build a magnet in which only coils, and not the poles, determine the field shape, but this would require very high currents and power dissipation due to losses (as given by Ohm's law) in the copper windings. On the other hand, a coil made of superconducting material has no resistance and will pass huge current without any ohmic loss. Once established, the current will flow forever. Of course the technology is somewhat more complicated than it sounds. The coil must be wound to a very pre-cise shape if the field is to be pure and uniform; and the magnetic forces trying to blow the magnet apart are huge. Each magnet has to be mounted within a "dewar," a huge cryostat, to keep its temperature below the crit-ical temperature for superconductivity, which, for the superconductors used, is only a few degrees above abso-lute zero.

The development of a method of embedding the fine superconducting filaments, some only ten microns in diameter, into a copper wire that could be woven into a cable, was crucial to building practical superconducting magnets.

Martin Wilson at the Rutherford laboratory in the UK was instrumental in developing a cable that could withstand the pulsing of a synchrotron while Alvin Tollestrup and Helen Edwards solved the challenge of using superconductors for the first superconducting synchrotron: the Energy Doubler at Fermilab (Fig. 5.13).

Fig. 5.13 Fermilab's superconducting Tevatron can just be seen below the red and blue room temperature magnets of the 400 GeV main ring. The Tevatron, unlike later superconducting synchrotrons, had a cold coil but a warm yoke. (In HERA and LHC both yoke and coils lie within a single cryostat.)

Martin Wilson (1939–)

British physicist
Developed the first practical superconducting cable
Developed solenoids for the first MRI scanners
American Physical Society R.R. Wilson Prize, 1989

Martin Wilson was born in Doncaster; his father was a bank clerk and his mother a school teacher. He has one younger brother, born after the war, who has become a professor of astronomy. His undergraduate work was at the University of Manchester, and although he obtained a "first," and was urged to stay on for a PhD, he elected to go into practical work with the UK Atomic Energy Commission. There he worked on gas-cooled reactors. Much later in life he collected his many publications and submitted them to the University of Manchester, which then granted him a DSc degree.

In the 1960s he joined Peter Smith's group at the Rutherford Laboratory, next to Harwell, in the UK. The group's task was to produce higher fields for accelerator magnets and Peter was keen to use superconducting magnets. Unfortunately there were two major obstacles to be overcome. The first was a sort of instability, which afflicts all superconductors. The ideal superconductor dissipates no power at all when it carries a current, but in the rising magnetic field of a synchrotron there is an effect called "flux jumping" (a sort of avalanche in which energy stored by the magnetic field gets converted into heat) which dissipates power and degrades the performance; even causing loss of superconductivity altogether as it warms up. The second is a hysteresis loss suffered by all superconducting materials. Currents induced in the conductors try to oppose the change. The group realized that both problems could be solved by making the superconducting wire as a filamentary composite, in which very fine filaments of superconductor are embedded in a matrix of copper.

For a large synchrotron, a high operating current is needed, requiring many superconducting wires to be connected in parallel. To make sure the current is shared equally among the strands of wire it must be twisted, or rather woven or braided like a good quality shoelace. The particular configuration developed by Martin and his team became known as the Rutherford Cable and has since been used in all superconducting synchrotrons.

Hearing of these developments, Oxford Instruments, not far from the Rutherford Laboratory, asked him to develop the first large solenoids for MRI scanners, which have become standard diagnostic equipment for large hospitals worldwide. Grafted onto a healthy industrial root-stock, Martin's enthusiasm for accelerators then led them to develop Helios, a storage ring x-ray source powerful enough to produce microchips by x-ray lithography, but compact enough to be transported intact by road, and more recently, Oscar is a 12 MeV PET-Isotope cyclotron weighing only one-tenth as much as a conventional machine.

Martin Wilson is a good example of how inventors of a new technology who intended to extend the capability of particle accelerators, can move from the accelerator laboratory into industry and find an application that is profitable and benefits the general public.

Alvin W. Tollestrup (1924–)

American physicist
American Physical Society R. R. Wilson Prize, 1989
National Medal of Technology, 1989
Distinguished Alumni Award from California Institute of Technology, 1992
Member of the National Academy of Sciences, 1996

Tollestrup was born in 1924 in Los Angeles, California. He obtained a BS from the University of Utah in 1944 and a PhD from the California Institute of Technology in 1950. He was a Research Fellow at the California Institute of Technology from 1950–1953, becoming a full professor in 1962. In 1977 he moved to the Fermi National Accelerator Laboratory, where he remained for the rest of his career.

At Caltech, after graduating in low-energy nuclear physics, he helped construct the 500 MeV synchrotron (originally an electron model for the Berkeley Bevatron). In 1957 he spent a year at CERN, where he did experimental work, for the first experiment, on their cyclotron. Upon returning to Caltech he was involved in experimental work on their synchrotron and, later, in the design of a 300 MeV proton synchrotron. This was never built, but the thinking led to the very large machines at Fermilab and CERN.

At Fermilab, besides the contribution to superconducting magnet technology described in the text, he played an instrumental role in detector development for the Tevatron Collider. In fact he was co-spokesperson for the design, construction and operation of one of the two detectors at the Tevatron. In addition he was very involved in the development of neutrino factories and muon colliders, a grassroots effort involving physicists from around the world, brought together with the motivation to accelerate muons. Muons have a lifetime of only 2 micro-seconds, so one must "move along" rather quickly and many of the usual techniques used in accelerators cannot be employed. Thus the production, "cooling," and acceleration of muons presents many challenges to accelerator physicists — challenges being addressed at this very time.

Helen Edwards (1936–)

American accelerator physicist
MacArthur Foundation Award and Fellowship, 1988
Department of Energy E.O. Lawrence Award, 1986
Fellow of the American Physical Society
National Medal of Technology, 1989
Member of the American Academy of Arts and Sciences
Member of the National Academy of Engineering
American Physical Society R.R. Wilson Prize, 2003

Helen Edwards was born in Detroit, Michigan, and received a BA from Cornell University in 1957. Her Master's degree and PhD came from the same University in 1963 and 1966. She remained at Cornell as a research associate until 1970, at which time she moved to Fermilab where she remained.

Helen Edwards was introduced to accelerators while a graduate student at Cornell University. As a postdoc, under Bob Wilson and Boyce McDaniel's direction she led commissioning of the Cornell 10 GeV Synchrotron. When Bob Wilson became director of Fermilab he recruited her to join his team and help build the laboratory. It was Edwards who was to be responsible for commissioning the fast cycling Booster, which continued as the workhorse of Fermilab.

She was very active in getting the Fermilab machines and their extraction system to work during the difficult period of 1970–72 and spent many long nights in the control room battling with their teething problems. Although diffident in social gatherings, she showed formidable strength of purpose in her professional activities and let no one stand in her way.

She learned from the early Fermilab experience that the next project, the Tevatron, would have to be built to higher standards of reliability if the technology of superconducting magnets were to be implemented on this large scale. A strong team of Rich Orr, Dick Lundy, Alvin Tollestrup and Edwards was assembled to see to it that this was done.

She was part of the initial management team for the SSC and spent two years there, departing in early 1991 after the project initiation phase.

Following the success of the Tevatron, she was invited to DESY to help Bjoern Wiik with his vision of a dramatic improvement in superconducting acceleration through the formation of the TESLA collaboration. This connection brought superconducting RF technology to Fermilab. The 2004 selection of the superconducting approach for an International Linear Collider attests to the success of Wiik's endeavor and Edward's foresight.

Martin Wilson's cabling activities led directly to a valuable spin-off to the medical field, where superconducting cable was used to build the solenoids that are the principal component of the Magnetic Resonance Imaging (MRI) whole body scanners, now to be found in most large hospitals in developed countries.

Chapter VI. Colliders

VI.1 Science Motivation

There are two kinds of high-energy particle physicists — theorists and experimenters. Theorists seek mathematical descriptions of elementary particles and the strong, weak and electromagnetic forces through which they interact. As the description improves it will, more often than not, predict that a new particle should exist, if only for a brief instant, and it is the experimenter's job to find it and measure its properties in order to test the validity of the theory. There may be more subtle predictions that give credence to the theory, but the discovery of a new particle is the sort of challenge an experimenter enjoys.

To find the new particle the experimenter will fire a high-energy proton or electron at another proton or electron (or their antiparticles) and look to see if the new particle emerges among the debris of the interaction. The impact may take place when a beam of particles hits a particle at rest (fixed target experiments) or when it meets another beam of particles head-on in a collider.

Sometimes the experimenter can sift through the data from old experiments to see if the new particle has been missed, but the most common reason for not having seen it is that the new particle was not there; the energies in use did not equate, in Einstein's formula, to so massive a particle.

The remedy is simple, but expensive: build a new accelerator to accelerate the projectiles to a high enough energy to create the new massive particle.

One is also faced with the choice of which particles to collide. Protons or antiprotons are an obvious choice; however, these turn out to be complex objects. Each consists of three quarks held together by gluons. In collision between hadrons — the class of particles that includes protons and antiprotons — only one quark in each of the colliding hadrons is involved in the interaction. The two quarks that collide can at most only carry one-third of each proton's energy; in fact the situation is worse than that, for each quark carries not a full third, but only about a fifth, of the hadron's energy — the rest is stored in the gluon field.

The physicist's job of analyzing the collision is not made any easier by the fact that quarks have different properties called flavors and one can never be sure of the flavor of the particular quarks involved in the collision. Nevertheless hadrons — protons and antiprotons — are unique probes for the study of the strong interactions.

Alternative projectiles, electrons or positrons, are simpler objects than hadrons. Rather than being made up of multiple quarks, these "leptons" are more or less irreducible, pointlike objects. They are especially well suited for studies that involve electromagnetic forces. The entire electron's energy is available to make new particles. In a recent neck to neck search for the next new particle, the Higgs, the 100-GeV electron-positron collisions in CERN's LEP proved a worthy competitor to a 1000-GeV proton-antiproton collider at Fermilab. In fact, neither machine found the Higgs, and it is now probable that this particle will only appear in the data from CERN's soon-to-be-completed LHC — a proton machine with 7 TeV per beam.

There are also experiments made by colliding protons, heavy ions, and electrons and even proposed experiments colliding beams of high-energy gamma rays. All contribute to a full picture of the forces that govern the interaction of particles.

VI.2 Principles

In the early years of particle physics, experimenters examined the secondary particles produced when high-energy protons or electrons from a synchrotron struck protons or neutrons in a fixed target. This was usually a simple rod of copper or other dense metal in the path of the beam extracted from the accelerator. Today's experiments probing the frontiers of particle physics examine the fragments thrown out when two beams of particles collide head on. The modern preference for head-on collisions is because they provide more energy for new particle creation.

Imagine for the moment a more familiar kind of collision — that between automobiles. In a road accident the damage caused is a measure of how much energy of motion (kinetic energy) has been released. A car glancing off the guard rail continues on its way without losing too much velocity or kinetic energy and, provided it does not hit any other traffic, the damage may be superficial. Contrast this with two cars meeting head-on. The cars and their debris end up with little or no velocity and all their kinetic energy is converted to bend metal and wrench the bodywork apart. The same is true of particles, but the energy is so high and the damage so complete that out of the wreckage come new particles. Einstein's rule equating mass and energy tells us that the more energy that is available, the more massive are the particles that are likely to be reconstructed from the debris. Head-on collisions, whether they are between automobiles or subatomic particles, make available all the kinetic energy of the projectiles.

Colliders were slow to appear on the stage of particle physics but this was not because people were unaware of their advantages. Kjell Johnsen, one of the pioneers of large storage rings, remembers that even prior to World War II, when accelerators were in their infancy, good physics students knew that the "available energy" from a collision of a sphere with a fixed object only increases very slowly with kinetic energy compared with head on collisions between two spheres.

The experimenter would obviously like to see as many interesting events emerging from the collider as from a beam hitting a fixed target. One point in favor of the storage ring collider is that the beams are continuously colliding, not just once every few seconds when a burst of particles emerges from a synchrotron. But this is by far not enough to make up for the low density of the colliding beam compared to a solid target.

The real obstacle to be overcome before a collider could be built was to produce a circulating beam of sufficient density to have a finite chance of producing a new and interesting "event." Without improvement in density it was like colliding two clouds together and it took many years for accelerator technology to develop ways to make adequate density. Most particularly, the effects of space charge had to be overcome. There are many aspects of space charge ranging from direct effects to wake effects due to wall resistivity. (See

Albert Josef Hofmann (1933–)

Swiss physicist
American Physical Society R.R. Wilson Prize, 1996

Albert Hofmann is probably the most widely liked personality of the accelerator field. He greets everyone with a courteous interest in what they are about and is hard to provoke into a controversial discussion. If he wants to put an opposing view it will often be hidden behind a funny anecdote or quotation from the life and work of some distinguished physicist — often Pauli or Einstein. With contacts all over the world and a wealth of experience from helping other laboratories to design and improve their accelerators, he is frequently asked to lecture on his specialties — beam instabilities and diagnostics. At schools and workshops his measured explanation of the more knotty points in this most mathematical of technologies is invariably appreciated by the beginners and professionals in his audience — many of whom may have struggled long and hard to understand the tortuous explanations of less gifted experts.

After his graduation at the Swiss Federal Institute of Technology ETH, he moved to the Cambridge Electron Accelerator (CEA), where he worked on the conversion of the synchrotron into an electron-positron colliding beam facility and carried out some particle physics experiments. CEA closed in 1973, and he moved to CERN with the title of senior physicist. There he worked on accelerator physics problems at the Intersecting Storage Rings (ISR), an important hadron collider of its day, then on the design of the Large Electron Positron (LEP) collider. In 1983 he accepted a joint chair in applied research at the Stanford Linear Accelerator Center (SLAC) and the Stanford Synchrotron Radiation Laboratory (SSRL). Here he worked mostly on the development of the Stanford Linear Collider (SLC) and on accelerator issues related to the production of synchrotron radiation.

He returned to CERN in 1987, and was jointly responsible for the commissioning of LEP. Following its completion, he worked on accelerator physics problems of this machine until his retirement in 1998. After, he worked for shorter periods at SSRL, the SLAC B-factory (PEP-II), the Brazilian light source (LNLS) and recently for Lyncean Technologies, Inc., Palo Alto.

With such a breadth of experience, and with so many friends in the accelerator community, he is a natural candidate for the machine advisory committees that keep every big project on course; and his wise but moderate counsel has aided many of them.

sidebar for Space Charge in Chapter III.) Many physicists have worked on this problem; dozens upon dozens have devoted their whole professional careers to the subject. We single out Albert Hofmann, but there could be very many.

Nowadays we calculate the likelihood of seeing something interesting from a collider by using a figure of merit called the luminosity. The luminosity is proportional to the product of the number of particles in one beam and the density of the particles in the other. The number of particles is limited by collective beam instabilities, while the density — at least for proton machines — is determined when the beam is injected into the machine and when space charge repulsion causes the beam to widen until it is stable. The first effort to reduce beam size by focusing was by Gus Voss and Ken Robinson.

Luminosity

Luminosity is an index of performance for a collider. Let us imagine for a moment that we are riding on the back of a particle in one of the beams and we collide with an oncoming cloud of particles. Each oncoming particle appears like a black disc with an area that we call the cross-section of a particle. If there are N such discs and the total area of the oncoming beam is A then the total black area is N times the cross section of each particle. The probability of the collision is then the total obscured area divided by A, the area occupied by the oncoming beam. We must then multiply this probability by the number of other particles with which we are traveling and the number of times per second that the beams see each other. This gives us a probability of an encounter, but it depends on the particle's cross section.

Physicists think of different kinds of collision events as having different cross sections. When the event, like the production of a rare new particle, is improbable we take this into account by imagining that the cross section of the oncoming particle is very small. On the other hand if the event is rather common, such as a small deflection or energy loss of the particles involved, the black disc seems large. If a new particle with rare properties is to be made the disc seems very small indeed.

Thus the performance of a collider expressed as a probability depends on the kind of event. We could of course measure the performance of a collider in terms of this probability for a particular kind of process, but we can play the trick of making its performance independent of the rarity of the event by simply dividing it by the cross section. This gives us a figure of merit which we call the "luminosity." Such a definition has the virtue that is depends only on the design of the collider and will be the same however rare or common the event we are seeking. It may seem an obscure mathematical trick but it is no different from dividing miles by gallons when we want to compare the performance of automobiles whatever the distance we have traveled.

Kenneth Robinson (1925–1979)

American theoretical accelerator physicist

Robinson was born in San Diego and did his undergraduate work and beginning graduate work at Caltech where he received a BS in 1948 and an MS in 1948. He moved to Princeton for his PhD, which was awarded in 1955 for experimental work in high-energy physics. From 1948–1952 he had been a research engineer at the RCA Laboratories where he worked on scintillation counters, transistors and color television.

In 1955 he was perhaps the first, certainly among the first, of those recruited by Stanley Livingston to the Cambridge Electron Accelerator (CEA), which was just being built. Robinson remained at the CEA until it was shut down in 1974. He then retired, moved to San Diego, and stopped communicating with his many colleagues and friends in physics. Sadly, he died only a few years later of a heart attack in his apartment, to be buried by the Public Administrator.

At CEA Robinson studied the effect on electrons of radiation while the particles were being held in a storage ring by means of external RF. He put forward, for the first time, a full analysis of the subject known as "radiation damping" and produced a theorem still known as the Robinson Theorem. He independently invented the Free Electron Laser in an unpublished work. With Gus Voss he invented a "low beta" concept for focusing beams to very small transverse size, an essential technique still used today for all colliders.

Robinson was a retiring and modest person. He was an avid swimmer and, also, an extensive walker. Very quiet and not one to easily make friends; he never married. His physics, however, was innovative, accurate, and highly respected by his colleagues. He was a "loner" and a "genius." Although he was a person who rarely spoke, when he did it was beautiful to hear; clear, logical, economical, precise and always correct. Gus Voss recounts a story which illustrated this:

During the design and construction of the CEA the director Stan Livingston had weekly staff meetings. In the course of one of those, a new idea was brought up which people thought would be a considerable improvement to the design of the synchrotron. Livingston immediately grabbed the idea. During the meeting he started to design magnets and many others joined in with further suggestions, such that after a considerable amount of brainstorming, everybody felt very happy and proud and leaned back — rather content with themselves. Suddenly Livingston's eyes fell on Ken, and he realized that Ken had not taken part in this pow-wow. Livingston said, "Ken, you did not say anything. How do you like the idea?" Robinson replied, "Well, it won't work."

Everybody seemed to be shocked and crestfallen. Nobody dared to open his mouth. Finally the director asked, "Ken, why won't it work?" After thinking quietly Ken gave a precise one-sentence answer, which needed a few minutes of quiet digestion on the side of the director and the staff, but then it was quite clear he was right. Livingston said, "Ken, you sat here all the time. Why didn't you say a word?" Robinson replied, "You didn't ask me."

In the late 1960s the time was ripe to build a practical storage ring. Sharing the honors of being pioneers are Bruno Touschek, Donald Kerst, and Gersh Budker and the laboratories at which it happened were Stanford, MURA, the Cambridge Electron Accelerator, Orsay, Frascati, CERN, and Novosibirsk. Kerst was instrumental in the development of proton-proton colliders, while Touschek and Budker spearheaded the development of electron-positron colliders.

VI.3 Electron–Electron Colliders

The first of these was the Stanford-Princeton machine built by G.K. O'Neill, W.C. Barber, B. Richter and W.K.H. Panofsky — all of whom in later years became major players on the stage of accelerator construction. This proposal was based upon a letter by D.W. Kerst *et al.* that had stimulated D.B. Lichtenberg, R.G. Newton and H.M. Ross, and independently O'Neill to propose making the collisions in a storage ring. (See Fig. 6.1.)

Although the Stanford-Princeton machine was the first collider to reach the drawing board, Kerst's idea of storing and colliding had not gone unnoticed in Russia at Novosibirsk where Budker and Skrinsky fashioned the rival to the Stanford-Princeton machine, called VEP1. (See sidebars for Budker and for Skrinsky in Chapter X.) The Novosibirsk Laboratory, under the direction first of Budker and then of Skrinsky, has made very many important contributions to high-energy, accelerator, and plasma physics. (See sidebar for Novosibirsk and see Fig. 6.2.)

VI.4 Electron–Positron Colliders

Positrons are the anti-particle of electrons and, although they have identical mass, their electric charges are opposite and equal — negative for electrons and positive for positrons. Suppose a collider magnet's polarity is set to bend negative current towards the centre of the ring. A beam of electrons circulating clockwise is seen by the magnets that guide it as a negative current. An electric current is just the product of the charge and its velocity, hence a beam of positive charges circulating anticlockwise will also appear as a negative current and be deflected to the centre of the ring. Both beams may circulate in opposite directions in the same storage ring and will collide together all round the circumference. Instead of two storage rings, one will suffice.

The same idea would work also for protons and antiprotons, though as we shall see, antiprotons are not so easy to make in large quantities as are positrons.

It was Touschek who built the first electron-positron collider, called AdA. (See sidebar for Touschek.) Although construction on AdA started later it was completed before the Stanford-Princeton electron-electron collider and started operation in Frascati near Rome in 1961. The source of positrons at Frascati was too weak

Gerard K. O'Neill (1927–1992)

American physicist

Gerard O'Neill was born in Brooklyn and went to Swarthmore College, from which he graduated in 1950. He received a PhD from Cornell in 1954, and from there moved directly to Princeton University where he remained, and was active until a few days before he died from leukemia.

O'Neill had three careers in his lifetime: as a high-energy physics experimentalist, as a teacher and writer advocating human colonies in orbit and on the moon, and as an entrepreneur who initiated a number of high tech companies.

In 1956, O'Neill invented storage rings. He was part of the team that made the very first electron-electron set of rings: work that was important in setting a direction which high-energy physics has followed ever since.

Apart from his activities as a high-energy physicist, O'Neill is widely known for his advocacy of manned space travel. This began in 1969 when, already a full professor, he still taught Princeton's elementary course in physics. He asked the class members to write a term paper on human habitation in space, and found the responses so affected him that until his death he devoted himself to various aspects of space habitation. He soon became internationally known as an author of a number of books on this subject and as spokesman for people around the world interested in space colonization. In 1978 he founded a privately funded Space Science Institute.

As an entrepreneur, O'Neill also founded companies involved in an inexpensive satellite navigation system, an office communication system and a high speed train system similar in configuration to a telephone network.

Burton Richter (1931–)

American physicist
Nobel Prize, 1976
Co-discoverer of the J-psi particle

Burton Richter was born in New York and entered MIT in 1948. There he had to decide between chemistry and physics but in his first year found physics more interesting. Particularly important teachers were Francis Bitter, with whom he worked as an undergraduate assistant after his 2nd year and Francis Friedman, who Richter says "opened his eyes to the beauty of the subject." In his third year Martin Deutsch aroused his first interest in the positron-electron system, which was to be the *leitmotiv* of his research throughout his career. As a graduate student he began studying the structure of the mercury isotopes in the MIT magnet laboratory under Francis Bitter. Some of the isotopes were radioactive with short lives and had to be made with the MIT Cyclotron. This was his first taste of the accelerators that, together with particle physics, were to become his lifelong passion. After a time at the Brookhaven Laboratory, where the 3 GeV Cosmotron was in its youth, he returned to MIT, working with Louis Osborne at the 300 MeV synchrotron, which students were expected to operate and maintain — an excellent training ground for what was to come. He then moved to Stanford where there was a 700 MeV linac: the precursor of the two-mile Stanford Linear Accelerator, later to play an important part in his fortunes.

Together with O'Neill, Barber and Gittelman he started work on the first electron-electron storage ring. Others may have claimed to have known of the idea earlier and other labs were to overtake the six year construction schedule with electron-positron colliders, but it was his team that first had the courage to start hardware construction of a device which was the intellectual ancestor of all storage rings.

Richter's approach to all new ideas was one of unremitting enthusiasm. Anyone coming into contact with this gregarious New Yorker would be treated to a pithy and unequivocal statement of the virtues of the latest theory, device or technique that was his current passion. It was this enthusiasm and the care he took, in dramatic pauses, to make sure that it was working its magic on the audience that won him a place in the folklore and affections of the accelerator community. His powers of persuasion placed him in the position to make his Nobel-winning discovery with the electron-positron collider ring, SPEAR, which was his pride and joy at SLAC.

His ideas were not to be constrained to the North American continent and his enthusiasm for electron-positron colliders and physics they produced, led him during a visit to CERN, to be among the founding fathers of the LEP project. Later he, together with Skrinsky of Novosibirsk and Tigner of Cornell, were the first to realize that the lepton machine to follow LEP would have to be a linear collider in order to avoid punitive synchrotron radiation energy losses. His attention switching to this new topology, he decided that the SLAC linac with a couple of arcs added for electrons and positrons could make 100 GeV collisions and scoop LEP. This most difficult and novel project was built before LEP and was within a hairsbreadth of preempting LEP's lead in electron-positron collisions when LEP's much higher luminosity prevailed.

Nevertheless SLC, still the only collider for polarized leptons, provided the practical basis for the linear collider studies which have dominated the project strategy of the high-energy physics community for a couple of decades.

Burton Richter continues as a powerful voice in the linear collider community advocating the techniques that SLAC has developed under his leadership as Director, from 1984 through 1999. As a flavor of his commitment one can only quote his own words that he has had, "… *a long love affair … with the electron. Like most love affairs, it has had its ups and downs, but … the joys have far outweighed the frustrations.*"

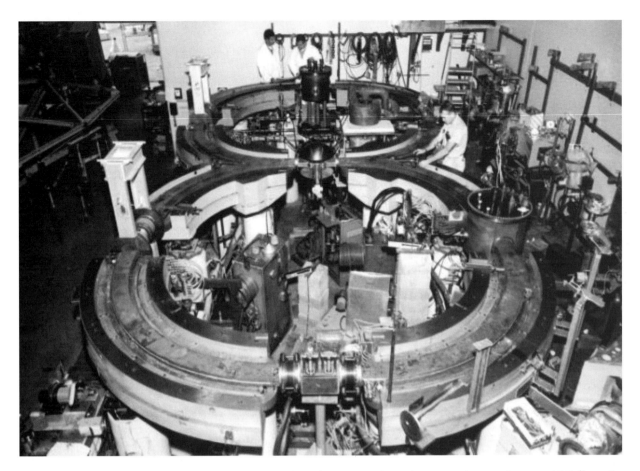

Fig. 6.1 The electron-electron storage rings, as seen in the early 1960s, at the High Energy Physics Laboratory (HEPL) on the Stanford campus. It took many years to make these rings able to produce physics and establish a much more stringent test on the validity of quantum electrodynamics. These rings provided invaluable experience for the further development of colliders.

Gersh Itskovich (Andrei Mikhailovich) Budker (1918–1977)

Russian physicist
Corresponding member of the Russian Academy of Sciences, 1958 and Academician, 1964
USSR State Prize, 1951, Lenin Prize, 1967 and Order of the October Revolution, 1975

Budker was born in a village, son of a rural laborer. In 1936 he joined Moscow State University, from which he graduated in 1941. He immediately became an anti-aircraft gunner and developed his first invention: an improvement of the gun guidance system. After the war, he worked on mainly plasma physics at the Kurchatov Institute, (1946–1954) and as an instructor at Moscow State University (1947–1949).

While at Dubna he developed an interest in accelerators. He investigated resonances, shimming of magnetic fields, beam extraction and proposed some radical ideas at the first international accelerator conference in 1956.

Budker developed the idea of mirror machines with end plugs for fusion energy. With eight associates at the Kurchatov Institute he started work on electron storage rings. In 1958 his activities were moved to Novosibirsk - Akademgorodok and the Novosibirsk State University where he built his Institute, part of the Siberian Branch of the Academy of Sciences. Under Budker's leadership, this became one of the greatest institutes in the Soviet Union. His team of thousands built one electron storage ring after another, pursued high-energy physics, built commercial accelerators, and investigated the mirror machine approach to controlled fusion.

In 1966 he invented beam cooling — particularly cooling of hadron beams with electrons (see sidebar on Cooling).

Budker characterized himself as a "relativistic engineer" but, he was really more than that. He never joined the Communist party, and ran things in a democratic way, holding a daily "round table" where all were equal. His idea to sell accelerators brought money into the Institute, gave it worldwide attention and led him to be accused of being a "Capitalist." Years later, when the Soviet Union collapsed, the new director, Sasha Skrinsky, was accused of being "Socialist," when he shared the income over all the divisions of the Institute. The Institute had not changed — only the regime.

In the course of his lifetime Budker had five wives and, as one of his colleagues remarked, "Budker does everything in a big way."

The Budker Institute of Nuclear Physics at Novosibirsk

Renamed in honor of its founder after his death, the Budker Institute of Nuclear Physics (BINP) in Siberia at Novosibirsk has made an outstanding contribution to the field of particle accelerators. At the same time it has been an important player in the field of fusion power, a subject beyond the scope of this book. It grew out of the Laboratory of New Accelerator Methods in the Institute of Atomic Physics, the Kurchatov Institute, in Moscow. The move to a bare site in Siberia took place in 1958 and the laboratory, then called the Institute of Nuclear Physics (INP), immediately began to generate a number of extremely advanced ideas for acceleration.

In the frigid years of the Cold War physicists from the INP would brave the uncertainties of immigration procedures to attend conferences in the US and Europe to explain, in often halting English, a whole nursery of brain-children. It is in the nature of such original ideas that only a few can be expected to bear fruit, yet this laboratory, thanks to the inspired leadership of Gersh Budker and Sasha Skrinsky, has seen more than its share survive to maturity.

The first major contribution it made to the field was the development of the 2×160 MeV electron-electron collider VEP1. The conception, design and operation of this machine closely shadowed that of the first US collider, the Princeton Stanford machine. Started in 1961, VEP1 was producing physics results in 1965, and it was followed within a couple of years by the electron-positron single-ring collider VEPP-2 (of 2×700 MeV). By 1970 this too was producing copious physics results. An even higher luminosity machine, VEPP-2M, reached a luminosity of 10^{30} in 1975.

One of Budker's original idea (1965) had been to use a beam of electrons, traveling with an antiproton beam to soak up the excess energy of rogue antiprotons and "cool" the hadron beam. This was worked on theoretically and experimentally with a model proton-antiproton ring, NAPP-M, and first demonstrated in 1974. It was about this time that proton-antiproton colliders became a realistic proposition, and electron cooling is quickly taken up and studied at CERN and Fermilab as a rival to the stochastic cooling that was finally adopted. Electron cooling is faster at high antiproton or ion beam intensity than its stochastic alternative and works much better at low energy. It is therefore used today in a number of low energy ion and antiproton rings.

Many of BINP's innovations have depended upon excellent engineering for their execution. Typical of such ideas is the use of a current of several thousand amperes in a bar of lithium to focus antiprotons produced after a high intensity proton beam hits a metal target. The lithium, which explodes on contact with water, must be nevertheless be cooled (indirectly) with water. To master such technology in a highly radioactive environment is one of the most difficult engineering feats in the accelerator field, yet BINP engineers were the first to do so.

BINP has for a long time had a parallel program of mass producing electrostatic accelerators — mainly for grain sterilization. This program has helped to keep it going in the difficult times of the breakup of the old Soviet Union. In another effort to bring prosperity, BINP was at one time under contract to produce a large number of components for the SSC — but that machine, sadly, was never completed. However, they went on to make major contributions to LHC.

It has its own programs for an asymmetric electron positron collider: the phi factory, which, with typical BINP ingenuity, squeezes two superposed figure eight "rings" through a single detector to double the luminosity as forth and back beams clash. The electron and positron beams in each of the pairs of butterflies wings have different energies (hence the name "asymmetric"). This topology was perhaps inspired by the so-called "Siberian Snake," an interlaced collider orbit that preserved the polarization of the spin of the beam's particles in collision and which was an earlier BINP idea from Derbenev and Kouratenko.

BINP scientists are as always eager to contribute at the frontier of accelerator technology and they participate vigorously in plans for a high energy linear collider and a free electron laser.

BINP is a large laboratory with several thousand employees, which combines excellent theoreticians with some of the best engineering workshops in the world. It has never been driven to exclusion of accelerator development by the demands of its physics program. It has always been at pains to keep in touch with the latest ideas in the western world and taken pride in rapidly developing new ideas in advance of its western rivals. When the world's accelerator scientists gathered there in 1976 following an accelerator conference, there were banquet toasts to the many ideas that BINP had given the world — so many that the seemingly inexhaustible hospitality of the USSR on such occasions almost ran out — and that was thirty years ago.

---

and the machine was taken to Orsay where there was a positron linac and where collisions were sufficiently frequent to do particle physics. Unlike the Frascati configuration, in Orsay the ring, once loaded with electrons, was flipped over like a pancake to inject positrons in the opposite sense using the same linac. At first this did not work until it was realized that in the flipping process, dust in the vacuum system had been stirred up into a cloud that obstructed the beam. (See Fig. 6.3.)

With the successful operation of the Stanford electron-electron rings, and the success in Europe with electron-positron rings, large electron-positron storage rings could be constructed with confidence. The Frascati

Laboratory went on to build ADONE (see Fig. 6.4). Then one saw the progression of SPEAR and PEP at Stanford, whose construction was led by John Rees, and DORIS and PETRA in Europe, at DESY in Germany, whose construction was led by Gus Voss (see sidebar for Voss). The DESY Laboratory has contributed in many different ways to high-energy physics, as has been commented upon throughout this book. (See sidebar for DESY.) A very important leader of this laboratory (and there have been a number of fine physicists in that position) was Bjorn Wiik. (See sidebar for Wiik.) In the early 1970s the Cambridge Electron Accelerator (CEA) in Cambridge, Massachusetts actually turned their syn-

Fig. 6.2 The first Soviet electron-electron storage ring, VEP1. Construction was started in Moscow, moved to Novosibirsk in 1962, and as soon as 1965 was giving results on electron-electron scattering. There are two rings, each of 43 cm radius. The equivalent "fixed target" energy was 100 GeV.

Fig. 6.3 The first electron-positron storage ring, AdA. Built and operated at Frascati, Italy, it was later moved to take advantage of a more powerful source of positrons in France.

Bruno Touschek (1921–1977)

Austrian physicist
Builder of AdA — the first collider for physics experiments

The development of accelerators largely overlaps the most turbulent period in European history, when Nazi racism expelled many gifted scientists from their home country, causing them to flee, dislocating their careers and the projects in which they might be engaged. In the US, the outbreak of World War II put a full stop to the building of cyclotrons as acute minds were harnessed to building spectrometers, calutrons, for the electromagnetic separation of fissile material for weapons.

Bruno Touschek was one of those not fortunate enough to leave his country in time. His native Austria, the birthplace of Hitler, was preparing for its annexation by Germany. In 1937, while still at school in Vienna and, in spite of being top of his course, he was asked to leave because of Jewish blood in his ancestry. Hearing of this, Arnold Sommerfeld (of quantum theory fame) invited him to Hamburg, where his origins were less well known. He was befriended by Wilhelm Lenz, in whose house he lived, and by Egerer, who gave him employment working in the offices of the *Archiv für Electrotechnik*. It was here that he learned of Wideroe's plan to build a 15 MeV betatron and by 1944 he was hard at work using the rigorous mathematical rules of Hamilton to solve the problem of stability of the betatron's circulating beam orbit.

By then he moved to Berlin but when the end of the war was in sight he tried to escape the bombing by driving to the station in an electric cart carrying his precious library of books and catching the train to Hamburg. In Hamburg he immediately fell under the suspicion of the SS who could not understand his frequent visits to the chamber of commerce there, where he was able to read foreign newspapers. The SS arrested him, but had to move him from one prison to another; he escaped but collapsed carrying his books, was shot in the head, left for dead, but again turned in to the SS by the doctor who treated his wound.

It was during this period in prison that he developed the theory of synchrotron radiation, written in invisible ink in the margins of a quantum mechanics textbook and smuggled out by his friend Wideroe.

Touschek was a gifted theorist and during one of Wideroe's visits famously tried to discourage Wideroe's plans to collide beams. He thought these trivial and predicted that they would never be patented. Later however, Touschek, was to reconsider this advice and this led him to build the first ever collider.

Liberated by the Allies, he went to Glasgow where for five years he worked with J.C. Gunn on the 350 MeV synchrotron. It was during this time, and when lodgings were hard to find, that he is said to have manfully intervened to stop the landlord battering his wife to death. He also had a desk made whose drawers had no sides or backs (to avoid the collection of dust). Such an idea was ill suited to Touschek's impatient and fiery nature and very soon all the contents of the upper drawers fell to the lowermost level in a hopeless jumble. Other manifestations of his rather eccentric but imaginative nature were that he had two cats, "Planck" and "Pauli." One he claimed to be rather intelligent but the other was consigned to be less so, when he failed to open a tin of sardines even when given the key.

In 1952 the most fruitful period of his career started when he moved to Rome. At first he could only speak a mixture of modern English and ancient Latin but he clearly fell in love with the bohemian lifestyle and relied upon his motor-bike "Josephine" to find the way home after many a late night's revelry.

He worked on quantum field theory with Forreti and at the same time built AdA, the first collider to be used for physics. To complete this he took the whole machine to Orsay, Paris, where there was an electron linac capable of creating enough positrons to make the counter rotating beam. However, to load the machine with the positrons the whole device had to be flipped over vertically. At first the device no longer worked and it took some time to realize that a small amount of dust fell down during this manipulation and destroyed the stored beam.

He was viewed with great affection by his Italian and other colleagues. However his health began gradually to decline due to his lifestyle. In 1977, when he had just been appointed to a senior position at CERN, and was working on stochastic cooling, he succumbed to a terminal bout of nephritis.

---

chrotron into a storage ring. They designed a "bypass" that switched the orbiting beams of low energy into a section of the synchrotron on a parallel path to the synchrotron itself. It was incredibly complex. Operational by 1973, it would eventually collide e⁻e⁻ beams of 2.5 GeV each.

The most recent stage of development is represented by LEP, the large electron-positron ring recently decommissioned at CERN, and by the B-meson "factories" at SLAC and KEK (Japan). We shall say more about LEP and the B-factories later.

As we shall see in the discussion of LEP, extending the concept of circular electron-positron colliders to a higher energy involves the consumption of huge amounts of power and requires extensive real estate to compensate for synchrotron radiation losses. A solution proposed in the 1980s by Maury Tigner and Burton Richter consists of two linear accelerators facing each other. This machine, SLC — short for SLAC Linear Collider — used the two-mile long 50 GeV linac to accelerate positron bunches closely followed by electron bunches. The beams were guided around opposite semicircular arcs to collide in a single detector as shown in Chapter III, Fig. 3.6. Work on SLC started in 1983 under Richter with the aim of simply seeing if a linear collider could be made to work. Later it was to develop

Fig. 6.4 ADONE, the first of the large electron-positron storage rings. The figure shows the ring during construction; operation commenced in 1969.

John Robert Rees (1930–)

American physicist
Fellow of the American Physical Society

John Rees was born in Indiana. After obtaining his PhD in 1956 at Indiana University, where he had received his AB and MS degrees, he became a Research Fellow in Physics at Harvard University and was also a Guest of the Department of Physics, MIT from 1956 to 1965. In 1965 he became a staff member and, later, an Adjunct Professor at the Stanford Linear Accelerator Center (SLAC). At SLAC he was Associate Director 1980–1991, and then Emeritus Professor from 1994. He took leave of absence from SLAC (1967–1969) and worked as Chief of The Advanced Accelerator Branch of the Atomic Energy Commission (overseeing the National Accelerator Laboratory, later re-named as Fermilab). He also took leave from SLAC (1991–1993) while he was Project Manager and Associate Director, for the ill-fated Superconducting Super Collider.

During his Cambridge, Massachusetts years, he worked at the Cambridge Electron Accelerator (CEA), where he engineered and built the RF accelerating system using the conceptual design of Ken Robison. After the accelerator was completed, he participated in elastic scattering experiments on protons and edited the CEA 3-GeV storage ring proposal.

At SLAC he joined Burt Richter's "Group C" and played a key role in the construction of SPEAR. His accelerator knowledge, and management expertise especially on large accelerator projects, was soon appreciated and he was made Director of the PEP Project (1976–80) and of the SLAC Linear Collider (1983–86). After his term at the SSC he served as advisor to the SLAC Director during the construction of the B-Factory, PEP II.

To many of his professional colleagues John Rees is a quiet embodiment of the SLAC spirit of unhurried excellence. With his steady gaze and well-trimmed moustache he would not be out of place as one of the "white hats" on the set of a western movie. John Rees married in 1956 and is the father of two children.

DESY

DESY was founded in 1959 by the German government in partnership with the City of Hamburg. Its first assignment was to build a 7 GeV rapid cycling electron synchrotron. Its first director — Willibald Jentschke, a warm-hearted and charismatic Austrian — gathered around him a team of young engineers and scientists who, though at first inexperienced in accelerators, were quick to draw upon the expertise of other colleagues — notably those from the Cambridge Electron Accelerator in the US — and set themselves high standards for the synchrotron construction. Their first machine was completed within the first three years, staking their claim to be at the forefront of electron accelerators in Europe, complementing CERN who were, until the days of LEP, to specialize in the proton option.

The next DESY project was the construction of a 5 GeV electron-positron storage ring called DORIS. This was started in 1970 just before CERN's ISR was completed and the first experiments with DORIS began in 1974. During this time the European Laboratory for Molecular Biology became interested in the synchrotron light from DESY's electron machines — an interest that would lead in 1980 to the establishment of HASYLAB: a synchrotron radiation laboratory using a fair fraction of the running time of DORIS and later PETRA. DORIS became famous for its experiments revealing charmonium and opened up the physics of charmed quarks. It also saw the first tests of x-ray lithography with the aim of constructing high-density microcircuits.

Those were the days of the hunt for the top quark. No one knew its mass and hence the energy at which a collider might be expected to create it, but 20 GeV seemed a reasonable bet, and so the next collider PETRA was constructed (again on a three-year timescale). This was quite a big ring (one-third of the circumference of CERN's SPS) and it took some ingenuity to squeeze it into the DESY site. The project was led by Gus Voss whose project management was a balanced compromise between the cost-cutting style of Bob Wilson and the conservative approach of John

Adams. He imbued his team with enthusiasm and at one time they were designing the next machine, HERA, in the day shift while trying to get PETRA to work on the night shift. The PETRA control room was among the first to use modern computers and colored TV screens to display a polyglot mixture of German and English legends. Although the "Old School" German formality had been discarded to the point that no one wore a coat and tie, the Project Leader was still "Herr Voss" to his German staff and "Gus" to the other component of the Anglo-Saxon equation.

The 1980s saw DESY move into the international arena with an ambitious proton-electron storage ring on the same scale as the Tevatron. One of the rings carried electrons at the modest energy that PETRA would provide, while the other, the proton ring, used the superconducting magnet technology then being developed on a large scale at Fermilab. The mastermind behind this project was Bjorn Wiik, a gently spoken Norwegian particle physicist who became interested in the machines themselves and was from then on to lead DESY — first in the superconducting age of HERA and later that of the linear collider, TESLA.

Funding these large projects was beyond even what were then the ample resources of the West German economy, and DESY became an international laboratory with contributions from European labs and from overseas. There was therefore and still is considerable rivalry between DESY and CERN, which was not softened by the decision of CERN to go ahead with LEP, its first essay in the lepton collider world hitherto thought of as the intellectual property of DESY.

TESLA — a superconducting version of the "linear collider to be" — was DESY's next contender in the race to high energy. There is already a 33-km track of land mapped out for it in the fields of Schleswig-Holstein. For the moment its proponents have been encouraged to limit their activities to using their advanced technology — the first stage of such a machine, to build a free-electron laser, but, no doubt, they hope that when the site selection for a world collider takes place, their aspirations will be rewarded.

---

into a race to scoop LEP in colliding 50 GeV leptons and producing intermediate vector bosons before LEP. They found that tuning up this novel machine was very tricky; it took longer than expected and, although they produced a few events before LEP, they were almost immediately overtaken by LEP's much higher luminosity and therefore data production rate. Their achievement, however, was that they succeeded in making the first linear collider: an invaluable investment for the future.

LEP has roughly four times the diameter of CERN's 400 GeV proton synchrotron, SPS, but has only one-third of its energy per beam (see Fig. 6.5). The reason for this is that electrons, two thousand times less massive than protons, are closer to the speed of light than a proton of the same energy. Any charge circulating in an accelerator will emit electromagnetic radiation, but when its velocity approaches that of light, the loss

of energy by the particle multiplied by the number of particles in the colliding beams rises steeply and represents megawatts of power, which must be replaced from the electricity supply via the RF cavities. This can be reduced by bending with a large and gentle radius — hence the size of LEP. The whole topic of this synchrotron radiation, as it is called, is discussed in Chapter VIII. In fact, the very construction of electron-positron colliders depends upon a careful analysis of synchrotron radiation and most particularly upon analysis of the effect of the radiation back upon the radiating particles. The subject of the effect upon the particles was first described by Mathew Sands in 1969. (See sidebar for Sands.)

Herwig Schopper (see sidebar for Schopper) played a leading role in the design of LEP and in securing its approval. Schopper then became the Director General of CERN, appointing Emilio Picasso to lead the construction phase. (See sidebar for Picasso.) LEP will

Bjorn Wiik (1937–1999)

Norwegian physicist
Led the design and construction of DESY's HERA and their
contribution to TESLA

Bjorn Wiik was one of a handful of particle physicists who ended up leading the large accelerator projects of their day. After growing up in Norway and studying at Darmstadt, Bjorn Wiik did research at Stanford University and SLAC for several years before moving to DESY and the University of Hamburg.

In 1967, as a recent graduate and during his time at SLAC, he had the idea of building a new type of accelerator to collide highly energetic electrons with protons. The time was then not ripe for such an idea but many years later it re-emerged as the HERA project.

In 1972 Wiik came to DESY, Hamburg, and in 1979, he was one of the team of four physicists that found the first experimental proof of the existence of the gluon at DESY's PETRA (Positron-Electron Tandem Ring Accelerator). It was then that he again took up the electron-proton idea and together with accelerator experts from DESY and CERN led a full scale design study of such a machine, called HERA. He was particularly skillful in recruiting experts to put the flesh on the bones of his idea and in no time at all was able to mold the many disparate and conflicting components of the design into a coherent whole.

His style of leadership was that of a patient persuader, reforming the ideas of his colleagues to suit the needs of the project. Expressing his views with a gently lilting voice that, had he been Irish rather than Norwegian, might have been called a brogue, he would set adventurous goals to provoke his more conservative colleagues into proposing ingenious and more practical alternatives. The first design of HERA's electron ring consisted of a ring of magnets with three legs, the third resting on the support of its neighbour and all powered by a single embracing ring of copper bus-bar. Fortunately this set of dominoes was abandoned in favor of a more conventional ring, but not before the idea of economy had been firmly planted in the designers' minds.

There was also political acumen in his repertoire of skills, for he was able to convince the German government to rapidly embark on a project whose scale and use of superconducting technology had yet to be matched even at an international laboratory the size of CERN.

With the success of HERA behind him he became head of DESY in 1993 and led the laboratory steadfastly towards yet another ambitious project TESLA: a superconducting linear collider that will probably become the basis of the next particle accelerator on a global scale. For a time, it looked as if Germany had once again been persuaded to build a huge, advanced project close to Hamburg. If not for a sudden tragedy in which Wiik fell from a tree in his garden, this might well have come to pass.

Maury Tigner (1937–)

American physicist
American Physical Society R.R. Wilson Prize, 2000
Fellow of the American Academy of Arts and Sciences
Member of the National Academy of Sciences

Born in New York, he obtained his BS from Rensselaer Polytechnical Institute and his PhD from Cornell in 1963. After that, he stayed at Cornell as a post-doc, but soon rose to professor (1977). In the 1970s Tigner was the first scientist to bring the idea of building an electron-positron collider to Cornell following conversations with Bjørn Wiik at Hamburg's DESY laboratory. As project leader he was responsible for dreaming up a "fiendishly clever" method of filling the ring with positrons by coalescing bunches from the synchrotron. In 1979, the Cornell Electron Storage Ring (CESR) collided its first beams using this scheme and the facility ran this way for several years. It was the first step in a proud tradition at Cornell that would see CESR hold the world luminosity record for many years.

In 1994 he retired from Cornell and spent six years at the Institute for High Energy Physics, Beijing. Since the People's Republic of China opened its doors in 1978, western accelerator experts have been eager to take the opportunity to help this Institute establish what is now a vigorous tradition in accelerators. Tigner's prolonged stay was at a time when they were developing an upgrade to their Beijing electron positron collider (BEPC).

However in 2000 he was called back to Cornell and was appointed Director of the Laboratory of Nuclear Studies.

During his "retirement" he, and Alexander Chao, edited a comprehensive *Accelerator Handbook* that has already enjoyed a third printing. The book sits on the desk of essentially every accelerator physicist and engineer.

He enjoys a well-deserved reputation of being one of the world's most highly respected accelerator physicists because of his design expertise and leadership of complex scientific projects. He was called upon to lead the Design Study for the Superconducting Super Collider (SSC) and later became involved in the Next Linear Collider leading a technical committee examining design options. Tigner is a person of even temperament and non-controversial approach that serves him well in his various administrative roles.

Fig. 6.5 An overview of the European high energy physics laboratory CERN. Superimposed on the photograph is an outline of the ring (deep underground, of course) of LEP, this tunnel later to be used for the Large Hadron Collider (LHC) described later in this chapter. In the background are Lake Geneva, the airport and Mont Blanc.

almost certainly remain the largest electron-positron storage ring of all time. Its tunnel, 27 km in circumference, stretches beneath the landscape between the Jura Mountains behind CERN and the Geneva airport. It was first built to accelerate lepton beams of 50 GeV and even at this energy the circulating beams radiate megawatts of synchrotron radiation — radiated power that has to be restored to the beam by the RF accelerating cavities even when the beams are merely coasting at a fixed energy. The magnets do not have to be very strong to keep beams of modest energies circulating in such a large ring. In fact their yokes, which normally are built of massive stacks of closely packed laminations of high quality steel, were built of only a few widely spaced steel sheets filled in between with a very fine concrete rather like the mortar used to build brick walls.

VI.5 Superconducting Cavities

Readers may have been wondering why the cavities that accelerate particles have hardly been mentioned so far in connection with synchrotrons and colliders. The truth is that for proton machines the main problem has been to design magnets and the tunnel to house them in order to bend the particles so that they return to the accelerating cavities in the shortest circumference.

Electron rings need more powerful cavities to compensate radiation loss. Once LEP had been running at 50 GeV for a year or two the time came to upgrade the energy to 100 GeV per beam. The magnets' strength had been chosen with this in mind but the original RF cavities made of copper for use at room temperature were completely unsuitable for higher fields, as they wasted 90% of their power in ohmic losses in the walls. A more efficient cavity that could provide a higher voltage per meter was needed. Although a large fraction of the ring had been left empty to house future cavities the new cavities would have to provide a higher accelerating gradient, greater than 5 MV per meter length, to make up for synchrotron radiation losses at 100 GeV. Special niobium clad superconducting cavities had to be developed (a more complete description of the development of superconducting cavities is given in the sidebar entitled "Superconducting Accelerator Cavities," with a typical installation shown in Fig. 6.6 and Fig. 10.4 showing an example of the shape of a typical niobium cavity). This proved a huge task but the new technology was eventually mastered and LEP went on to run at full energy for the rest of its life.

VI.6 Proton–Proton Colliders

Returning now to the earlier days of colliders, we find that proton colliders were not slow to follow their electron counterparts. Proton-proton colliders, unlike electron and positron colliders, do not have the strong radiation damping effect to help them produce the very intense beams necessary for high luminosity. Instead they accumulate intense beams by a process called stacking.

Stacking was a product of work at MURA where Keith Symon and Donald Kerst had pioneered the fixed-field alternating gradient (FFAG) machine (Chapter II). This suggested to Donald Kerst the possibility of accelerating and storing a sufficiently intense beam to make a collider. The MURA group then developed the idea of stacking up several beams side by side. Their "two-way" FFAG was used to experimentally study and demonstrate the reality of such stacking. Despite these major contributions MURA was dissolved in the early 1960s without ever having received support to build a high-energy accelerator (see sidebar for MURA in Chapter II).

The work at MURA had attracted the attention of CERN scientists who had then just finished their 28 GeV Proton Synchrotron and were looking for a project to extend the energy reach of their machine. There were two options — scaling their "PS" up by a factor of 10 to be a 300 GeV synchrotron or taking the imaginative but risky step of building a proton-proton collider fed by the PS.

It made sense to use CERN's new 28 GeV synchrotron to feed two other rings (the Intersecting Storage Rings or ISR) dedicated to accumulating and colliding the beams. Stacking or accumulating beam is not a trivial process. Each turn as it is injected must not encroach on the space occupied by the beam that is already there or the injection process will simply drive the old beam out of the machine. The turns must be laid side by side differing slightly in energy. Before each turn is injected, a rather weak accelerating cavity nudges the circulating beam to a slightly lower energy to leave room for the incoming beam. It was planned to raise the ISR current in each ring to 30 A in this way.

It was realized that two concentric storage rings, one slightly square and the other diamond in shape could intersect in four collision points. One beam would circulate clockwise and the other anticlockwise and serve more than one experiment. The idea could be extended to make eight crossings.

Each intersecting ring was half as big again as the CERN PS or the AGS, then the world's largest synchrotrons. It was an adventurous machine to build and it

Mathew Sands (1919–)

American physicist
Fellow of the American Physical Society
American Physical Society R.R. Wilson Prize, 1998
Distinguished Service Award of the American Association for the Advancement of Science, 1972

Mathew Sands was born in Massachusetts, and obtained a BA from Clark University, followed by an MA from Rice Institute in 1941. He was at the US Naval Ordnance Lab from 1941 to 1943, and then at Los Alamos from 1943 to 1946. After the war, Sands was at MIT from 1946 to 1950 where he received his PhD in 1948. He then joined the Caltech faculty, where he remained from 1950–1963. From 1963–69 he was Deputy Director of SLAC. In 1969 he moved to the new campus of the University of California at Santa Cruz and remained associated with that campus, as professor of physics, chairman of the physics department, and vice-chancellor of the campus. His formal retirement was in 1985.

In 1952, during his time at Caltech, Sands went to Italy as a Fulbright Fellow and then later to France as a visiting professor at the University of Paris in 1976. He was president of Sands-Kidner Assoc. Inc. from 1986–1991. He has been a consultant to the Office of Science and Technology.

Sands has made a number of very important contributions to physics. He was the first person to realize that quantum effects could have macroscopic consequences on electron beams. His 1969 monograph on this and all other aspects of electron beams in storage rings is still read widely. It was he who in 1959 first suggested cascading accelerators to reach very high energy, a scheme now followed in all high-energy accelerator installations. While at SLAC he was very involved in initiating the effort on high-energy storage rings: work leading up to SPEAR.

He is the author or co-author of a number of technical books but is probably most widely known as co-author of the *Feynman Lectures on Physics* (1965).

Sand's theoretical ability in the field of accelerators combined with his technical skill in this field and in nuclear and cosmic ray physics has produced many imaginative contributions to the field of accelerator physics. These talents coupled with a fine administrative ability have enriched his leadership credentials.

Throughout his life Sands has been very active in the field of arms control. He was a consultant to the Institute of Defense Analysis and a member of the Pugwash Conference on Science and World Affairs as well as of the Federation of American Scientists.

was unclear whether it would work. This machine was designed to collide beams of 20 A (a staggering hundred times more than contemporary synchrotrons provided).

Herwig Schopper (1924–)

German physicist
Director of the Institute for Experimental Nuclear Physics of the
   Technische Hochschule and the Nuclear Research Centre (KFK)
   Karlsruhe, 1961–73
Director General of CERN, 1981–8
President of German Physical Society, 1992–4
President of European Physical Society, 1994–6
President of the SESAME Council, 1999–
AIP Tate Medal and UNESCO–Albert Einstein Gold Medal

Although now a German national, Herwig Schopper was born in 1924 in Landskron, Bohemia — now part of the Czech Republic. He received a diploma in physics in 1949 and a doctoral degree in 1951 from the University of Hamburg and subsequently worked as a Research Fellow at Stockholm with Lise Meitner, and the Cavendish Laboratory (see sidebar on this laboratory in Chapter I) with Otto R. Frisch, two pioneers in the study of nuclear fission. From 1960 to 1961 he worked at Cornell University with Robert R. Wilson (see sidebar for R.R. Wilson in Chapter V).

He tells the story, "When I arrived at Cornell in 1960 on a Saturday afternoon I went immediately to the accelerator hall. Nobody was there except somebody who was sweeping the floor. I started to talk to him, asked question about the machine and the physics done with it and was surprised about the intelligent answer I received. So finally I said: 'But you are not a janitor!?' 'No' was the reply, 'I am Bob Wilson.' When I expressed my surprise that he was cleaning the floor he told me that this was not unusual, when he came as a young scientist to Berkeley he had an analogous experience and the man sweeping the floor was Alvarez."

Returning to Europe, Schopper directed the Institute for Experimental Physics at the University of Mainz from 1957 to 1961 and was Director of the Institute for Experimental Nuclear Physics, Research Centre and the University of Karlsruhe from 1961 until 1973. In the 1950s, together with Clausnitzer and Fleischman at Hamburg and Erlangen, he developed the first source for polarised protons, later to be installed at the cyclotron in Karlsruhe. At Karlsruhe he set up a group to investigate the feasibility of superconducting rf cavities. At first the experimental cavities were coated with lead but later niobium became the preferred material. These were the first attempts in Europe to use superconducting cavities. The group was reinforced by Herbert Lengler from CERN, leading to a collaboration with CERN, which resulted in the first use of superconducting cavities for beam separation, first at CERN and later at Serpukhov.

At Karlsruhe he also started a group to develop the Electron Ring Accelerator (ERA) or 'smokatron' to trap protons and accelerate within a ring of electrons but like everywhere else who worked on the idea, these attempts failed and were never published.

Karlsruhe also then had the intention to build a proton synchrotron for 30 GeV but these plans were of course shelved when the decision to build the SPS was taken. Superconducting magnet research still continues at Karlsruhe but now mainly for the development and testing of magnets for fusion.

From 1973 to 1980 Schopper oversaw the Deutsches Elektronen Synchrotron (see sidebar for DESY): the main accelerator facility in Germany.

Although best known as a distinguished nuclear and particle physicist Herwig Schopper has contributed to many areas in physics, including optics, thin metal layers, elementary particle detector development, and accelerator technology, writing and editing many books on these subjects.

He has provided leadership worldwide in the creation and use of high technology research facilities in particle physics. While still heading DESY, which was then the pre-eminent electron accelerator laboratory in Europe; it was natural that he should become involved in CERN's plans to construct their first major electron collider, LEP. He was eager to fashion this project in the "no-frills" style he had learned to admire while working with R.R. Wilson and which he and others had implemented with great success at DESY in the construction of PETRA. He was therefore the natural choice for CERN's Director General from 1981 to 1988, a period of office that was extended to cover both the construction and completion of LEP. From this position he was able to introduce radical economies to ensure the huge project was built within tight budgetary constraints and on time.

He is remembered as a strong-minded Director General, greatly respected by the Member States and always able to find time to consider the views of all who might contribute to the efficient running of the Laboratory. He was able to effectively reconcile the many demands on his time from international contacts and committees, along with the practical business of running CERN, with the minimum of administrative backup.

His contribution to international scientific cooperation has been enormous. After his term of office as Director General of CERN he was President of the European Physical Society, and then from 1993 to 2002 he was a member of the Scientific Council of the Joint Institute for Nuclear Research (JINR) in Dubna, Russia.

Since his retirement from laboratory management he has dedicated himself to promoting physics in small and developing countries and in particular to the SESAME initiative (see sidebar on SESAME). The idea of SESAME is to bring together recently (and currently) hostile countries in the Eastern Mediterranean to rebuild a synchrotron light source in Jordan, much as the countries of Europe forgot their post-world war differences in the establishment of CERN. Schopper has been very successful in persuading UNESCO to take on this project — again in the image of CERN. He is the President of the UNESCO SESAME Council and is a member of the Founding Committee for the Research University at Cyprus.

To avoid scattering off residual gas the beam chambers had to be baked to a temperature higher than most domestic ovens. This was to drive away water vapor and other surface layers. It was, at that time, the largest ultra-high vacuum system in the world.

At the time that CERN embarked on ISR no one had ever really tested whether proton beams, which unlike electron beams do not experience the beneficial damping effects that come from giving off synchrotron radiation, would circulate for many hours. Various other

Emilio Picasso (1927–)

Italian physicist
Project leader of CERN's LEP collider
Légion d'Honneur (France)
Knight of the Grand Cross (Italy)

Emilio Picasso was born in Genoa and then studied at the University of Genoa. At first he studied mathematics, but after two years he switched to physics. After obtaining a PhD degree he became assistant professor in the chair of experimental physics. He started research in atomic physics and later on in elementary particle physics. For the latter, he worked at the betatron of Torino University and, later, at the electron synchrotron in Frascati. In 1961–1962 he worked on cosmic rays with the group of Prof. Powell in Bristol. In 1963 Picasso came to CERN to measure the anomalous magnetic moment of the muon.

Emilio Picasso says he has never been able to make up his mind whether he is a research physicist or an engineer — he enjoys both professions, and at many times in his career has moved from one field to the other. In his early days he was an expert on the quantum mechanics of a spinning particle — the muon — and its "anomalous" magnetic moment. He was recruited to join a team which included John Bailey, Francis Farley, Simon van der Meer, Guido Petrucci and Frank Krienen who, following a suggestion from Leon Lederman, were determined to see just how much small deviations from quantum electrodynamics might affect the muon's electromagnetic moment. Two muon storage rings were built at CERN and over 15 years the deviation was measured to be 1162 parts per million to the phenomenal accuracy of 5 parts per million.

This work completed, he became interested in gravitational waves and built superconducting resonators which would detect relative changes in the amplitude of such waves as small as 10^{-21}. The fact that no such waves have yet been directly detected down to this level is itself an intriguing result for physics.

In mid 1980 Emilio Picasso became project leader of LEP (the largest electron synchrotron and collider in the world). Until then, the design work of LEP had been overseen by Kees Zilverschoon, Eberhard Keil and Wolfgang Schnell and, as Picasso saw it, such a project was very much like an orchestra of expert musicians needing a conductor to ensure they make music out of their efforts.

During its construction, John Adams, then CERN's Director General, took him to one side to ask him, almost as if it was an idea that had just occurred to him, "Emilio, why not put all that superconducting technology that you have developed for gravitational wave detectors to good use in building LEP." At that time LEP was about to embark on the building of its first stage (50 GeV per beam). It turned out that superconducting RF was the key to doubling this energy.

His skills as a Project Leader were quite unique. He saw no place for the tantrums and bad temper of the symphony orchestra conductor and embarked upon every discussion with the members of his team as if he were their greatest friend — arm around their shoulder — asking them a favor or sharing an idea with them about what to do next — always full of that humor and boyish camaraderie that disarms divisive criticism and unites a team. Indeed, he had to use this skill to the full several times during LEP's construction, and it never failed.

At that time that he took over, LEP was an even larger ring on paper than it is today and stretched 1000 meters under the nearby Jura Mountains — a leaking reservoir of high pressure water waiting to inundate anyone with the temerity to tunnel into its foothills. Consulting a well known tunneling expert in Zurich he was told that he should move LEP to another place or suggest someone else should take the responsibility for what would end up as a project beyond the means of CERN's budget. He chose the former solution, reducing the radius of the ring, tilting it slightly to fit the contours and sliding it from under the mountains into the drier sandstone of the alluvial plane between the Jura and Geneva airport.

Although no one knew the energy needed to produce a pair of intermediate vector bosons, he was concerned to keep the high energy option of LEP up to 100 GeV — just high enough as it turned out. As an engineer he was also at pains to persuade his team to save money and to this end pushed Chris Benvenuti's novel money-saving idea of the distributed NEG pumping system. He was able to finally convince his team after a test which he persuaded DESY to perform.

In between and after his "engineering activities" he had found time to be Director of Nuclear Physics for CERN and finally, as he reached CERN's retirement age, he became director of his alma mater — the Scuola Normale Superiore di Pisa. He now continues his search for gravitational waves in collaboration with the University of Genoa.

horrible effects were expected. Many aspects of the collider concept were tested on a low-energy (5 MeV) electron model called CESAR. (See Fig. 6.7.)

Project director Kjell Johnsen and his team were fully aware of these dangers and insisted that the new storage rings should be conservatively built in all conventional aspects so that they might concentrate on the new and unpredictable effects. To do otherwise might have put the whole project at risk. Years later, Feynman in his report on the space mission Columbia disaster was to point out the dangers of compounding too many untried systems, a pitfall that Johnsen had wisely avoided. (See sidebar for Johnsen.)

His policy paid off, for at the end, some aspects of conventional nature had to be "pushed" far beyond their original design in order to make the collider work up to specs. The ISR was completed in 1969 and soon worked beautifully. It eventually reached 31.4 GeV, a current of 50 A in each ring and a luminosity of ten times the design value. (See Fig. 6.8.)

Superconducting Accelerating Cavities

Accelerating cavities are at the heart of linacs and, therefore, it has been the linac people who have been the leaders in developing ever better RF accelerating cavities. There was considerable work on S-band, L-band, side-coupled cavities, etc. at SLAC, Los Alamos, and many other places. At first these were room temperature cavities, but rather early on — when superconducting materials first became available — their attention turned to the use of these materials so as to reduce the power consumption and improve the duty cycle of their machines.

In the 1960s work on superconducting cavities was initiated at Lausanne, CERN, HEPL (Stanford), and Cornell. Although lead was first examined, very soon effort was focused upon niobium. A collaboration between Karlsruhe and CERN, initiated in 1970, produced a successful niobium cavity for a radio frequency particle separator by 1978. Meanwhile, in the US, the main thrust of research into the use of superconducting cavities was at HEPL at Stanford, where the first fully superconducting L-band linac was completed (after many years of very difficult R&D) in 1977. During the same time period the Argonne National Laboratory developed their superconducting linac. Research and development at Cornell also greatly advanced the art.

By the 1980s there was considerable activity in quite a large number of places. Working closely with Cornell, the Jefferson Laboratory built the re-circulating linac, CEBAF, with superconducting accelerating cavities. Early work at Wuppertal soon led to a very long and fruitful collaboration with DESY. Considerable advance was made, both with the type of cavity and very fine "packaging" into cryostats. This work led to TESLA, the DESY design of a TeV superconducting version of a linear collider that was, in fact, the design selected for the International Linear Collider.

Superconducting RF was, at first, of little interest to synchrotron builders as their cavities are usually of less importance in determining the size and cost of the machine than is the case with linacs. Indeed, in a journey round a large synchrotron one might easily miss the small fraction of the ring where the cavities are installed. The cavities do their work incrementally over many thousands of turns and for each turn a few MeV of acceleration is sufficient. However, their design and very type becomes much more critical for the larger electron storage rings where they must make up the energy radiated each turn. LEP was a prime example, originally built for 50 GeV; it was planned to double this to 100 GeV but at the cost of providing about 3 GV of accelerating voltage per turn to balance the losses. The normal accelerating gradient of a room temperature cavity is at best one or two MeV per m, so that a few km of the circumference would have to be filled with cavities, and then the power wasted in just heating up the copper vessels of the cavity would have been over 100 MW, a fair fraction of the output of a power station. Superconducting cavities were a necessity.

The modern room temperature cavity consists of a series of cylindrical structures joined together, with a central passage for the beam. These structures are of very special shape so as to reduce sparking and reduce the development of unwanted oscillation modes. The strong fields on the axis of these cavities are produced by oscillating radio waves not unlike the acoustic modes in an organ pipe. The waves end on the copper surfaces of the cavity — where strong currents flow to maintain the electric and magnetic fields of the waves. These currents dissipate thermal energy in the copper just as currents in a resistor — one cannot avoid the consequences of Ohm's law. However, just as in a resistor, the energy dissipated by these currents is proportional to the resistance: lower the surface resistance and the power loss is minimized. The ultimate — actually zero resistivity — is to coat the surface with a superconductor such as niobium. The incidental, but very important, by-product is that the voltage gradient can also be much improved in a superconducting cavity. This development did not come easily, but was the result of decades of R&D.

Early studies in the 1960s with lead coatings had revealed the problems of impurities and surface imperfections and led the technology in the direction of more exotic superconducting materials such as solid niobium or electrolytically deposited niobium. Any slight strain due to forming the cavity shape in a press or the seams due to welding the cells together would defeat the superconducting properties of the cavity, as would the slightest speck of dust on the surface layer. Nowadays, in large measure due to the R&D at Cornell and other places, superconducting cavities with large gradients and reliability of operation have been achieved.

Early RF cavities were incorporated into machines such as those at HEPL and ATLAS, and more sophisticated cavities were then developed for KEK (the first major installation of superconducting cavities in a circular machine; commissioned in 1987), CESR (Cornell), Jefferson Laboratory, LEP, HERA, and most recently TESLA. At the present time, superconducting RF is used both in linac and synchrotron-based colliders.

In the United States, design of a collider called ISABELLE was initiated just as the ISR was completed. After many years of frustration with superconducting magnets, and then with the authorities, this project was cancelled in the mid 80s even after civil engineering had been initiated and a tunnel for the machine had been constructed. (See sidebar for ISABELLE.)

Then, in the mid-80s the United States initiated work on a very large collider, the Superconducting Super Collider (SSC). After 10 years of design, actual construction of part of the tunnel, and untoward amount of work both by accelerator designers and particle physicists, the project was cancelled. (See sidebar for SSC.)

VI.7 Proton–Antiproton Colliders

After the ISR operated so successfully and even before LEP became an accepted project, Carlo Rubbia pointed out that the single ring of the SPS which was already up and running as a synchrotron might collide counter-rotating beams of protons and antiprotons. This procedure offered the next step in hadron energy for a minimal cost.

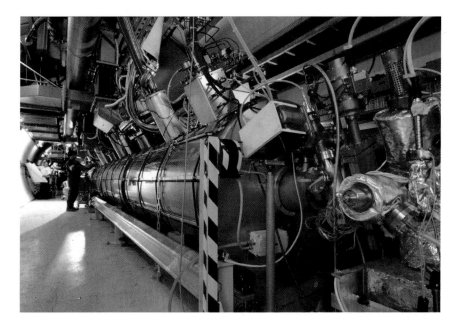

Fig. 6.6 Superconducting RF cavities at the CERN Large Electron Positron Collider (LEP).

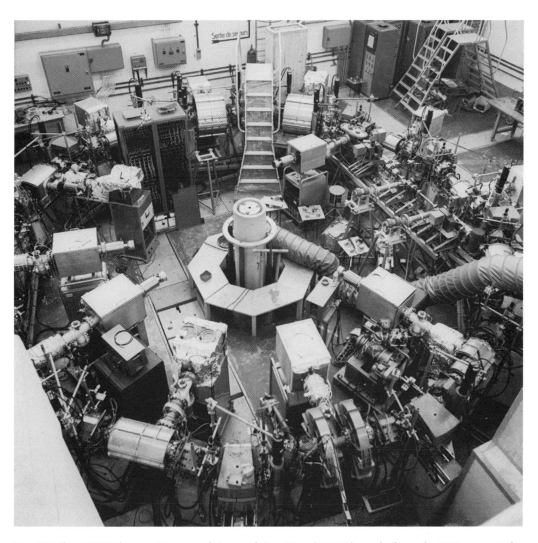

Fig. 6.7 The CERN Electron Storage and Accumulation Ring (CESAR) was built, in the 1960s, as a study-model for the ISR (Intersecting Storage Rings). All the stacking (accumulation) procedures envisaged for the ISR were proven with CESAR and critical aspects of transverse stability were explored. CESAR was also the test-bed for the ultrahigh vacuum techniques and components essential for the ISR.

Until then the supply of antiprotons had seemed too meager to make a respectably dense beam, but Rubbia had heard of two methods of concentrating the beam: electron cooling — proposed by Budker and Skrinsky and then demonstrated at Novosibirsk — and stochastic cooling, the brainchild of Simon van der Meer at the ISR. (See sidebars on van der Meer and Rubbia.) Rubbia tried to engage the interest of Robert Wilson, then director of Fermilab, whose 400 GeV ring could just as well be converted to collide p and $\bar{p}$ as CERN's SPS. However Bob Wilson was not impressed by Rubbia's idea and though Fermilab started some work on electron cooling, it was finally at CERN that Rubbia found the most enthusiastic support. An experiment, ICE, was quickly carried out using some old muon storage magnets. (See Fig. 6.9.)

Both cooling methods were tested and the stochastic method was chosen. (See sidebar for Stochastic Cooling and see Figs. 6.10 and 6.11.)

It takes a pulse of 10^{13} protons to produce the 10^6 antiprotons that can be collected in a reasonable beam channel, and it takes tens of thousands of pulses of protons to produce enough antiprotons for a proton-antiproton collider. However "cooling" made it possible for antiproton beams to be concentrated, slimmed, tapered, and shaped before being stacked in the vacuum chamber of a low energy (3.5 GeV) ring, where a seemingly endless supply of antiprotons could be accumulated. Antiprotons could then be harvested after many hours of accumulation, accelerated and then injected into the SPS to collide with an ordinary proton beam.

Fig. 6.8 The first proton-proton collider, the CERN Intersecting Storage Rings (ISR), during the 1970s. One can see the massive rings and one of the intersection points built with lots of room for collision detection devices. Although many beam dynamics issues had been tested on CESAR, many remained and it was by no means a certainty that the ISR would work as designed. At the end it did; and in fact, even exceeded its performance predictions.

Kjell Johnsen (1921–)

Accelerator physicist
Project leader of the CERN Intersecting Storage Rings
American Physical Society R.R. Wilson Prize, 1990

He graduated in 1948 with a degree in electrical engineering from the Technical University of Norway (Trondheim). While studying for his PhD, which was awarded by this University in 1954, he became a research assistant at the Chr. Michelson Institute Bergen. The Director of this Institute was Odd Dahl who was appointed to lead the Proton Synchrotron Group when CERN was formed in 1952. Dahl immediately enlisted Kjell's support. In 1952 Dahl, Goward and Wideroe visited the Cosmotron in the USA where they were told of the alternating gradient principle. Dahl switched plans for the European machine to take advantage of this revolutionary principle. Johnsen recalls how Dahl, on his return, asked everyone in the PS Group to "Drop all work and go straight onto this new idea. Forget about the report to CERN Council — I will see to that!" Kjell Johnsen moved with the PS Group to Geneva in 1953 to become the right-hand man of Hugh Hereward, then leading the Linac Group.

One of the problems was that the new synchrotron would have to cross a region known as transition where there was no radio frequency stability. It was Johnsen who showed how the beam would oscillate exactly like a variable mass pendulum and at transition would have infinite mass, preventing the growth of any instability. One trouble, however, was the effect, named after Johnsen, due to particles crossing transition at different times. He then developed the first successful radio frequency beam control loops, showing that one needed to add slow radial position feedback to fast phase control to achieve stability. He also invented the idea of a debuncher between linac and ring to match the longitudinal phase space.

When CERN's PS was finished he immediately turned his attention to colliding beams. His interest had been aroused by plans for FFAG machines at MURA, as reported in a 1956 Conference and later by G. O'Neill's opinion, "are not storage rings easier?" He was joined by Arnold Schoch and Kees Zilverschoon; and the accelerator research division which they led built a small storage ring (CESAR) to test stability and study stacking procedures. Kjell worked out the ISR design on the back of an envelope and the numbers hardly changed through the final design and construction. There were many skeptics who were against it but together they convinced CERN's Director General Viki Weisskopf to launch the ISR project in the face of considerable opposition. Even John Adams, a rival in some ways to Johnsen, admitted as he left CERN to head fusion research in the UK that "if necessary he would come back and build the ISR."

The section on proton colliders pays tribute to Kjell Johnsen's vision and persistence in completing this adventurous concept and surpassing its expected design performance.

Once the ISR was running and the SPS Project was in full swing, he led the first working group that together with the legendary Burton Richter generated the idea of LEP. It was Johnsen that invented the three letter mnemonic: LEP. He went on to look after the ISABELLE project at Brookhaven — an ill-fated superconducting ring that was later abandoned after he had returned to Europe but was to re-emerge as RHIC. He played an important role in the early days of the proton-antiproton collider and also CLIC, though modestly admits that the motive force for antiproton studies came from Carlo Rubbia and that it was Wolfgang Schnell who turned CLIC into a viable idea. (CLIC — the Compact Linear Collider, is discussed in Chapter X.)

One of Kjell's last and enduring professional contributions was as the founding director of the CERN Accelerator School which, after twenty years, is still helping to train the world's accelerator designers. He was also particularly pleased in his later career to be asked to serve as the Chair of the HERA Machine Committee at DESY.

The first antiproton accumulator, crucial to the whole idea, was built at CERN and enabled the 400 GeV SPS pulsed synchrotron to be converted into a proton-antiproton collider.

This was followed several years later by Fermilab which built a superconducting ring of 1 TeV (1000 GeV) nominal energy, the Tevatron, which soon was converted into, and still operates as, a proton-antiproton collider. (See Fig. 5.19 in Chapter V.) As at CERN the antiprotons are accumulated in special cooling rings (see Fig. 6.12).

VI.8 Asymmetric Collider Rings

Apart from purely proton and antiproton and purely electron and positron colliders, an electron-proton machine had been suggested in the early 1970s in the USA by a group that included some Europeans. The proposal was called PEP (which originally stood for Positron-Electron-Proton), but no support for it was forthcoming in the USA. (A positron-electron version was built, and later, its high-energy ring was used for a B-meson "factory," which we shall soon discuss in greater detail.) Like that other good American idea, the proton-proton collider, which was taken up in Europe to become the ISR, the electron-proton machine became HERA, built at DESY in Hamburg (see Fig. 6.13).

DESY built the two-ring system to collide 26 GeV electrons with 820 GeV protons. This may have been motivated by the fact that while electron storage rings (DORIS and PETRA) were a specialty of DESY, CERN had become interested in electrons for LEP. It seemed only right that DESY should do something with protons. This novel project, HERA, began operation in 1990. It

ISABELLE

The ISABELLE project had its origins in the heady days of the early 60s when the AGS and CERN's PS had come on stream after a neck and neck race to completion. Brookhaven's AGS was a little later but stole a march on CERN by having an experienced community of experimenters and equipment poised to take off. For a number of years the AGS at Brookhaven were leading the big-science race. Their success bred a certain complacency, so that first SLAC and then CERN began to catch up. There were moves afoot on both sides of the Atlantic to make the next machine a storage ring or collider and to look forward to a synchrotron of ten times the energy of AGS and PC.

CERN moved boldly forward with their plans for the ISR while the US decided to site a new 400 GeV synchrotron at Fermilab. Brookhaven, once at the top of the pyramid, now found themselves sidelined, but enlisted the support of East Coast friends in Washington to undertake to build the next big machine (a collider) at Brookhaven, establishing a pattern of rotating new projects among the three major accelerator labs.

It was John Blewett, veteran of the Cosmotron, who invented the ISA (Intersecting Storage Accelerator) and the Belle (from the French for beautiful), was quickly added. The project was to be 10 times the energy of the ISR and would need superconducting magnets to keep the beams on the Brookhaven site. Blewett hoped, in 1972, to have the project completed in time to celebrate the US bi-centennial in 1976.

Superconducting magnets were a very new technology and they soon became bogged down in a decade of unsuccessful R&D. So unpromising were the results that the DOE were persuaded to reverse their three-lab policy and award the next big slice of money to Leon Lederman at Fermilab to build the superconducting Tevatron.

Too late, BNL solved the magnet problem of ISABELLE or the CBA (Colliding Beam Accelerator as it had been renamed) when Bob Palmer adopted the kind of twisted cable pioneered by Martin Wilson and the Rutherford Lab. By then CERN had already discovered the W and Z with their antiproton collider and Fermilab was keen to make a p-bar source for the Tevatron — besides, the SSC was in the air. Attempts to revive ISABELLE's fortunes by bringing in first Paul Reardon to lead the magnet development, and then CERN's Kjell Johnsen with his recent ISR experience, could not prevent its demise in 1984. Fortunately Brookhaven's chance to recoup the many years of work and the embarrassingly empty tunnel was soon to come when RHIC was proposed as the next logical step in the field of relativistic heavy-ion physics pioneered by the Bevalac at Berkeley.

The Superconducting Super Collider (SSC)

At the centre of the Superconducting Super Collider Project, SSC, were a pair of synchrotron rings, designed to accelerate and collide opposing beams of protons and installed in a single tunnel 54 miles in circumference. It was intended to reach an energy of 20 TeV on 20 TeV: twenty times higher than that of the Tevatron, at Fermilab, then highest energy machine in the world, and forty times CERN's SPS. The project was first proposed in 1982 at a Snowmass Study, organised by the APS. At a cost of $2.9B to $3.2B, it was supposed to re-establish US supremacy in the field of high-energy particle physics following the discovery in Europe of the W and Z.

The next step was carried out. By 1986 a detailed design study, by a Central Design Group set up under Maury Tigner at Lawrence Berkeley Laboratory, was complete and in 1987 President Reagan set in motion the search for a site. In 1988 Waxahachie, Texas was announced as the successful candidate and construction commenced. This decision was, perhaps, influenced by then Vice-President George Bush (senior) of Texas, Jim Wright of Fort Worth, then Speaker of the House of Representatives, and a powerful Senator, Lloyd Bentsen, also from Texas.

Meanwhile the cost, taking into account the more realistic estimates of the Central Design Group, and including the proposed experimental facilities, had risen from the initial estimate of 2.9 B$ to 3.2 B$ and then to 5.3 B$ in 1986.

The Department of Energy took over the management of the project from the CDG and determined that the traditions of technical and cost control that had been built up in large laboratories, like Fermilab, were to be abandoned in favour of methods judged more appropriate for a project of this size. Contracts were placed and subsystems procured in a manner that had hitherto been used for large defense projects. This led to further escalation, first to 5.9 B$ and then, as review teams took into account site-specific costs, to 7.2 B$ and then 8.2 B$ (DOE, 1991). The final straw came when an Independent Cost Estimating Team of the DOE (1993) added 2.5 B$ for "peripheral expenses which would not have been incurred if the SSC had not been there." The SSC was given a year's reprieve but by 1994 — when the US Congress saw fit to cut their losses and terminate the project — the estimate was 11.8 B$.

The reasons for cancellation were not entirely budgetary. There was a rival project — the Space Station — that many in the House of Representatives preferred. There were, too, those who believed both projects should be sacrificed in order to balance the budget and it was also becoming intellectually fashionable to question whether science had at all enhanced the quality of human life. The impact on physics was much greater than at first appreciated; in fact the cancellation action had impact upon all of science for at least a decade.

Subsequently it was left to Europe to construct a more modest 7 TeV on 7 TeV collider, the CERN LHC, and to the Tevatron at Fermilab to hunt the Higgs, hoping that it might be found within its rather more limited reach in energy before completion of the LHC.

Simon van der Meer (1925–)

Dutch physicist
Inventor of stochastic cooling
Nobel Prize, 1984 (shared with C. Rubbia)

Simon van der Meer was born into a family of schoolteachers in The Hague in 1925. He was prevented from attending university by their closure under the German occupation and instead had to mark time in the humanities section of his High School and equipped his home with a number of electronic gadgets. Finally, after the war, he was able to study "Technical Physics," a subject well suited to his later activities in the field of accelerators.

He joined the recently-founded CERN in 1956 and, as CERN launched a project to make an intense neutrino beam to catch up with Brookhaven's AGS, he made his first original contribution to accelerator science, the pulsed device called the "neutrino horn," which focused the pi and mu particles, the parents of the neutrinos, in the direction of the experiment.

His next important contribution was his participation in the building of a small storage ring to store muons and measure their anomalous magnetic moment — so called "g minus 2" to great precision. Later when Cline, McIntyre, Mills, and Rubbia proposed to use either the Fermilab accelerator or the SPS for colliding protons and antiprotons and needed to store and concentrate the antiprotons, they remembered an idea he had had earlier called stochastic cooling. The idea had been known to be contrary to Liouville's conservation theorem — and was considered to be a heresy that could only work if a "Maxwell demon" were to put order into the random cloud of particles. Van der Meer has an imaginative mind which generates new ideas in abundance, but subjects them to a rigorous intellectual filtering process before announcing them as feasible. In response to Rubbia's proposal he re-examined the idea, found a means to make it work and participated in an experiment to cool a beam in a small ring: the Initial Cooling Experiment, ICE. He was able to show not only that stochastic cooling worked but that its rival electron-cooling, which also was tested in ICE was less suitable for antiproton cooling.

Although many Nobel prizewinners are cited as developing experimental techniques rather than discovering new phenomena there are few which are closely related to accelerators. Van der Meer's contribution to the Nobel Prize he shared with Carlo Rubbia was clearly one of these phenomena.

Unlike many of his Nobel colleagues, who almost invariably are propelled to great achievements by their self confidence, Van der Meer is a modest and quiet person preferring, now that he has retired, to leave the lecture tours to other more extrovert personalities, and look after his garden and occasionally see a few friends. Never has anyone been changed less by success.

Carlo Rubbia (1934–)

Italian physicist
Nobel Prize, 1984 (shared with S. van der Meer)
Discovered the intermediate vector bosons (with S. van der Meer)

Carlo Rubbia first began to study engineering at university when a sudden vacancy at the exclusive Scuola Normale, Pisa gave him the chance to change to physics. Thanks to this experience of both disciplines he was to retain a lifetime interest in both practical technique and the theory of science which colored and enriched his career.

From the time he joined CERN in 1960 it was clear to all that he was destined to make an important discovery in physics. His incisive intelligence marked him out even among the elite of the world of experimental particle physicists and his unwillingness to suffer fools gladly ensured that there was little danger of his remaining unnoticed.

His most impressive and fruitful period was during the gestation of the proton-antiproton project at CERN, which led him to the Nobel award for the co-discovery of the Intermediate Vector Boson. Turning the SPS into a collider would not have been possible without a means to accumulate and concentrate a large number of antiprotons. Budker at Novosibirsk had shown this might be done by a method of "cooling" using a co-axial beam of electrons. Van der Meer at CERN proposed a rival "stochastic" method of cooling developed in the ISR. Moreover it was not clear whether Fermilab's Energy Saver or CERN's SPS would be the machine to be converted to collide the particles. Dave Cline explored the Fermilab option while Carlo pursued the idea at CERN. Neither Laboratory seemed particularly keen on the prospect but it was John Adams, gently persuaded by his Co-Director General van Hove, who finally became reconciled to what Carlo might do to his brand new SPS machine. It was not clear how CERN's complex chain of accelerators might be best adapted to make cooling work and which of the two methods, electron or stochastic cooling, should be used. It was at this time that Carlo's genius for leading a team towards a coherent scientific proposal from a number of disparate ideas was most apparent. He was in his element contributing transcendental flashes of insight into the sometimes discouraging discussions of experts, cajoling them to accepting his optimistic view of the outcome.

Even at the height of his success, and afterwards as Director General of CERN, he generated a series of brilliant new ideas which he pursued in depth, but just up to the point that others might continue, realizing then that he had better move on to launch yet another imaginative leap. After the bosons, came fusion using ion beams to heat the thermonuclear pellets. This was followed by an accelerator-driven thorium reactor — the "Energy Amplifier," which doubled as a nuclear waste disposal facility. He later proposed a nuclear power source for space travel based on a critical reactor with direct fission fragment emission. All of these ideas were an intellectual marriage between technique and science.

Should he ever get to heaven, Carlo will no doubt replace the angels' wings with another means of levitation and then chastise them for not having the imagination to think of it themselves.

Fig. 6.9 In 1977 the magnets of the "g-2" experiment were modified in a record time of 9 months and used to build the proton-antiproton storage ring: ICE (the Initial Cooling Experiment). The ring verified the cooling methods to be used for the "Antiproton Project" and demonstrated to skeptics that the survival time for the antiproton was at least 32 hours: nine orders of magnitude longer than the previous best measurement of 2 microseconds.

Stochastic Cooling

Louiville's theorem seems to make it impossible to concentrate particle beams. Once a number of turns have been injected side by side in a ring they cannot be superimposed (see sidebar in Chapter X on cooling a beam). But early in the 1970s Simon van der Meer realized that it was apparently possible to beat this principle. He proposed using a pickup to measure the position of a particle at one point in a ring as it circulated and sending a signal across the diameter of a ring to activate a deflector kicker which would cancel any deviant behavior of the particle and put it back on axis. Of course the pickup could not measure the position of individual particles but rather sample the centre of charge of a number of particles, but statistics shows that by repeating the correction many times, rogue particles can be edged in the direction of the axis of the machine; this is called stochastic cooling.

(The word "stochastic" means to aim for a target and sometimes miss!) The process works best when the sample contains few particles. This is the case if the sample time is short which in turn implies working at as high a frequency as possible. The main difficulty is therefore technological: the pickups must be sensitive and, signal amplifiers must have both a broad bandwidth, high gain and be noise-free.

When developed, stochastic cooling proved to be remarkably effective and made possible the construction of the antiproton accumulator rings at CERN and Fermilab which fed proton-antiproton colliders. Antiprotons are produced in a very warm state, i.e., with a density which is too diffuse to give high luminosity. With stochastic cooling, the energy spread can be reduced by a factor of 100, and the transverse beam size also reduced by large factors — greatly enhancing the probability of collision.

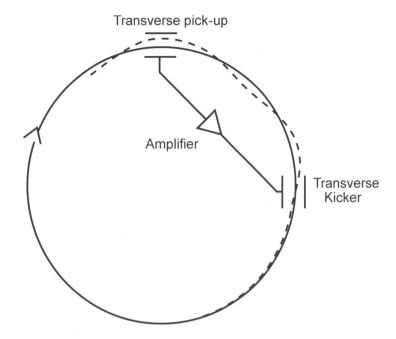

Fig. 6.10 Stochastic cooling works by using a transverse pickup to measure the displacement of a particle oscillating about the center line of the ring and sending an amplified signal to a deflector just as the particle arrives. The deflector steers the particle back on the center line. In practice, one can only measure the average position of many particles but statistically, it still works.

Fig. 6.11 A stochastic cooling pickup. Careful design and machining are necessary to make the electromagnetic sensor wave travel at the same speed as the beam and to have its amplitude slowly grow in response to the very small stochastic signal. The wave is then amplified and applied to the beam (see Fig. 6.10). The development of sensitive pick-ups and low-noise amplifiers is at the heart of practical stochastic cooling.

Fig. 6.12 The anti-proton, or "p-bar" source, built in the 1990s at Fermilab. The source consists of two rings that both accumulate the antiprotons and then stochastically cool them. The reduction in phase space density, the proper measure of the effectiveness of the cooling, is by more than a factor of 10^{11}.

Fig. 6.13 An aerial view of DESY in the city of Hamburg, Germany. Superimposed on the picture are some of the accelerators at this complex. In particular, the largest, the 6.3-km-long underground Hadron Electron Ring Accelerator (HERA) has been in operation for research since 1992. Almost all of HERA lies outside of the DESY site.

was the first European machine to use superconducting accelerator magnet technology for the complete proton ring and is unique in its ability to study the physics of high-energy hadron-lepton collisions — then a new and unique window on physics.

In addition to their interest in new and heavier particles some physicists spend their lives looking for violations of the laws of physics in the hope that this will give more insight into the laws themselves. One such law is that you can take a positive particle with right-handed "spin" and it will decay in exactly the same way as a negative particle with left-handed spin (all other properties being equal). Cronin and Fitch had already seen Charge-Spin (strictly charge-parity or CP) violation in K meson decays but when it was realized that it might also be observed in the decay of another meson, the B meson, there was enthusiasm to study this system. CP violation is believed to be the reason for the excess of matter over antimatter in the universe and is therefore responsible for our very existence.

The decay time of the B-meson is so short that the only way to see the violation is to make the B-mesons by colliding electron and positron beams of different energy. The centre of mass of the event then has a velocity in the direction the higher energy beam was traveling, and the decay time is revealed as a length of the particle track before it decays. This idea was due to LBL's Pier Oddone, who at the time of writing is the director of Fermilab.

Two great projects were dedicated to this kind of physics. One asymmetric pair of intersecting rings was built at SLAC, under the leadership of John Seeman. (See sidebar for Seeman.) The cross section for seeing B meson CP violation is very very small. The rings were designed for a phenomenal luminosity, but even this was surpassed.

At the same time at KEK near Tokyo, Japanese physicists have built a similar machine which has also performed "beautifully." Key members of the team that built the machine were Shin-ichi Kurokawa and Katsunobu Oide. These two machines have already established CP violation in B-meson decay and are now busily studying finer details of CP violation.

VI.9 Large Hadron Collider (LHC)

We have seen that often in the past, electron machines have been followed by proton colliders to chart higher energies. Now that LEP has done its job, the next piece of theory that needed to be verified once again predicts a very massive high-energy particle, called the "Higgs,"

John Seeman (1950–)

American physicist
Fellow of the American Physical Society
IEEE Particle Accelerator Science and Technology Award
American Physical Society R.R. Wilson Prize, 2004

John Seeman was born in Marshalltown, Iowa and received his BSc. degree in Physics from Iowa State University in 1973. His PhD from Cornell University was in accelerator physics; for designing and building the injection system for the electron-positron collider CESR. He spent the three years which followed his PhD as a research associate on CESR; studying injection, the vacuum system, and the beam-beam effect.

He moved to the Stanford Linear Accelerator Center (SLAC) in 1982 to help construct the SLAC Linear Collider (SLC), leading the Linac Group in its task of upgrading and operating this machine for the SLC. His particular concerns were in the domain of beam emittance control and wakefield studies.

In 1993 he joined the PEP-II B-Factory, becoming first the head of the High Energy Ring and later the Project Deputy for Accelerator Physics. He was the chief commissioner of PEP-II.

In 1999, once the construction of PEP-II was complete and his work in charge of commissioning the accelerator was over, he became the Head of the Accelerator Department at SLAC, which operates both PEP-II and the linear accelerator. Then, in 2002, he became an Assistant Director of the Technical Division.

Seeman is quiet, amiable, not easily ruffled and always works through consensus — qualities that have led him to serve on many committees including accelerator advisory committees at BNL, KEK, DESY, FNAL, IHEP, and Cornell, and also on several committees for the US Particle Accelerator Conference. He has taught at many Particle Accelerator Schools. His publication list includes over 150 articles. He, along with the PEP-II Management Team, received the 2000 US Department of Energy Program and Project Management Award for the PEP-II B-Factory Project. John Seeman, besides doing physics, enjoys hiking, photography, and gardening.

outside the range of existing colliders (see sidebars on the Standard Model and the search for the Higgs) (see Fig. 6.14). To find this new particle one needed a new hadron collider the "Large Hadron Collider" or LHC.

The sheer size and cost of this enterprise is without parallel and were it to be built on a "green field site" it is likely that it would have shared the fate of its higher energy predecessor, the SSC. The LHC replaces LEP in the 27 km tunnel; already excavated on the CERN site. It is fed by the chain of accelerators that already exist on the site. Nevertheless it is an ambitious machine for, in order to reach 7 TeV in the LEP tunnel, the mag-

Shin-ichi Kurokawa (1944–)

Japanese physicist
Head of the KEK Laboratory Accelerator Divisions
Constructed TRISTAN and KEKB

Shin-ichi Kurokawa, after his undergraduate course at the University of Tokyo, continued to receive first a Masters and then — in 1973 — a Doctor of Physics for his work on photo-production of pions by polarized gamma radiation on neutrons. He then started his career as an experimental high-energy physicist at KEK, constructing secondary kaon and antiproton beam lines for rare decay modes of the K-plus as well as the study of the antiproton annihilation at rest.

From 1981 to 1986 he moved away from the field of particle physics research to the Accelerator Department of KEK, where he participated in the construction of the control system of TRISTAN, a 30 GeV × 30 GeV electron-positron collider. His interest in TRISTAN deepened — first as its coordinator and then (from 1989 to 1994) as head of its Division.

He spent the next six years as project leader for KEKB: an asymmetric-energy, two-ring, electron-positron collider for B-physics. This project was commissioned in December 1998, and since physics experiments started in June 1999 its luminosity has been steadily improved, beating record after record and reaching 4.1×10^{33} cm^{-2}s^{-1} by June 2001.

His workload during this period was incredible even by the standards of dedication expected in Japan. By 2001 Kurokawa had become general head of the divisions of the KEK Accelerator Laboratory and was playing a leading role in the international relations of Japan with other laboratories worldwide, serving on advisory and review committees at SLAC, FNAL and BEPC-II in Beijing.

He is a great admirer of the Analects of Confucius and typically, as host of the 1996 Joint US, CERN, Japan, Russian accelerator school, opened the proceedings with some of the first words from the Analects: "Is it not delightful to have friends coming from distant quarters?" He had an excellent way of explaining the values of modern Japan to these "friends from distant quarters" in terms of the respect for education and learning his countrymen had inherited from the ancient civilization of China. He was and is particularly enthusiastic in keeping academic and scientific ties open with China and other countries in Asia. He was a prime mover in forming an Asian Accelerator School, first held in Beijing in 1999.

He has achieved remarkable success for his own country and in forming and sustaining friendship on a global scale. He is perhaps the best example of the latest generation of accelerator builders who face the challenge of building a World machine such as the International Linear Collider with a proven track record in accelerator technology coupled with a view that spans five continents and twenty-four time zones.

Katsunobu Oide (1952–)

Japanese physicist
Japanese Accelerator Promotion Prize and Nishikawa Prize, 1990
Nishina Prize, 2001
American Physical Society R.R. Wilson Prize, 2004

Born in Tokyo, he received his Bachelor of Pure and Applied Science from Tokyo University in 1975 and in 1977 his Master of Physics from the same university. In 1980 he also received a PhD from Tokyo University for research in the detection of gravity waves.

In 1981 he joined the accelerator division of the Japanese high energy physics laboratory KEK where he has remained ever since. At first he was involved with the construction and commissioning of TRISTAN, but soon became interested in computer code development which he applied to the design of the final focus system for linear colliders. He was a visiting physicist at SLAC 1988–1989, where he worked on the design of the Final Focus Test Beam used to study the making of the very small beams needed in linear colliders and for laser/plasma studies.

After a year at SLAC he became involved, throughout the 90s, with the design and construction of the KEKB B-Factory. At the same time (1990–1997) he worked on the design and commissioning of the KEK-ATF damping ring: a major component of the R&D necessary for a linear collider. In 1997 he was appointed head of the KEKB Commissioning Group and at the time of this writing is Head of the KEKB Accelerator Group. The pre-eminent world position of the KEKB B-Factory is in large measure due to Oide's work and the excellent group he has led.

Oide took on his first scientific project when he was in middle and high school, around 1967. He and his colleagues decided to build a computer out of about 400 used relays discarded by Japan Telegraph & Telephone Company, where his father was employed. A relay computer was already then out of date, but solid state circuits were too difficult for students. They designed a 12-bit computer with a system clock, an accumulator, an operation unit, and a few words of memory. They had difficulty connecting wires to relay pins without the proper tools, handling the dc power supply for the huge number of relays, and the more sophisticated problem of adjusting the timing of the switching without an oscilloscope. Nevertheless, the hands-on work was critical in Oide's development as it has been for almost all experimental scientists.

Oide likes to read books on European, Chinese and Japanese history and philosophy. He claims that although he does not understand, nor remember most of them, they are good for forgetting beam physics and inducing sleep with good dreams.

The Standard Model

The Standard Model describes four forces, transmitted by elementary particles: the photon transmits the electromagnetic force; the gluon carries the strong nuclear force that holds the atom's nucleus together, and the W and Z bosons transmit the weak force that acts in radioactive decay.

To explain two or more phenomena in a single theory has been the aim of particle physics for more than a century. The first success in this direction came with Maxwell's combined theory of electricity and magnetism. He found a way of describing these two forces of nature in the same equation. This led to countless practical applications — e.g., electric motors and generators and later the transmission of radio signals, TV and wireless phones — applications that are at the heart of our modern technology and now so familiar that we often forget that without physics they would not have been possible.

When it came to fully understanding Maxwell's unification of electrical and magnetic forces a crucial discovery was that the photon — seen alternatively as a particle or a wave-packet of light — was one of the actors on the stage. The photon, which appeared or disappeared whenever the forces acted upon matter, seem to "carry" or transmit the forces. Richard Feynman invented a cartoon or diagram in which the photon appeared as a wiggly line linking the change in direction of one charged particle with another as they collided. The rules of physics were completely described by these convenient diagrams, and this enabled physicists to predict what would happen to both particles after collision. The probability of various outcomes could be characterized by a coupling constant — or, in more complicated encounters, four or more constants arranged in a matrix.

For the best part of a century the force binding protons and neutrons in the nucleus remained a mystery, as did the weak force which manifests itself in radioactive — so-called beta — decay. Nevertheless the search was on for particles which might carry the strong and weak forces and for ways to describe the interaction with Feynman's diagrams and matrices. The discovery that protons and neutrons are themselves made of three quarks, each with different properties, seemed at first to complicate the quest, but at last a theory in which the carrier was a "gluon" seemed to make sense and explain the existence of so many related particles — including mesons — as different combinations of quarks. Pioneering work, in fact the introduction of "quarks," resulted in the Nobel Prize to Murray Gell-Mann in 1969. Later work, which made possible the description of strong interactions (called quantum chromodynamics), was done by David Gross, H. David Politzer and Frank Wilczek, for which they received the Nobel Prize in 2004.

The intermediate particle for the weak interactions of leptons which included both electrons and neutrinos was difficult to find but eventually the carrier particles — intermediate vector bosons, W and Z particles — were predicted by theory and created and identified.

To be precise, the theoretical unification of the electromagnetic force and the weak force was recognized in the Nobel Prize of 1979 to Sheldon Glashow, Abdus Salam and Steven Weinberg, Further work on this unification was done by Gerardus 't Hooft and Matinus Veltman who won the Nobel Prize in 1999. Experimental work at the CERN p-pbar Collider, confirming the theory, resulted in the Nobel Prize being given to Carlo Rubbia and Simon van de Meer in 1984. The whole unifying description of the forces of nature (save for that of gravity) fell into place and has stood up to a couple of decades of attempts to find flaws in it. We call it the Standard Model and as an intellectual advance it is surely one of the great triumphs of the last half of the twentieth century.

Gravity is yet another matter — a force that the standard model has not yet included and another loose end to be tidied up — is the reason why the fundamental particles of the standard model have the inertial masses that they do. Why, for instance, is the electron 2000 times lighter than the assembly of three quarks that we call the proton? Current theory seeks the explanation in the existence of yet a new — and heavier — particle: the "Higgs."

Just recently it has been observed that neutrinos have mass (very small mass, but non-zero mass). This is beyond the Standard Model and the very first indication that the model must be extended.

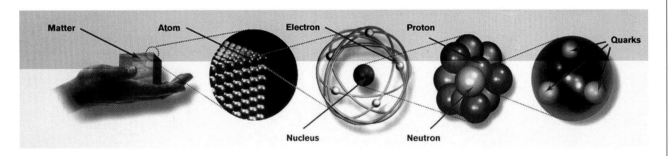

Fig. 6.14 An artist's view of how matter is made up of ever smaller constituents. As far as we know, the smallest point-like constituents are quarks and they cannot be further subdivided. The theoretical framework for this, called "The Standard Model," is perhaps the most important major advance in physics during the last 50 years.

netic field must be close to 9 tesla. This is five times the field at which a normal magnet would saturate and more than 50% higher than that for which previous accelerators such as the SSC had been designed. Only superconducting magnets can provide this (see sidebar in Chapter V).

LHC is a conventional two ring interlaced proton-proton collider like the ISR, but rather than two sepa-

The Search for the Higgs

Although the standard model describes most things about particles as different as electrons, protons and neutrinos, it cannot explain their seemingly unrelated masses. Physicists hope that the discovery of a new force-mediating particle — the Higgs boson — will add the "dimension of mass" to our understanding. In non-mathematical terms the reasoning reduces to whether the laws of physics can assume a certain symmetry: an important and unifying feature in any theory. Making the laws of dynamics symmetric, so that time and the three dimensions of space are on the same footing, was a crucial ingredient in Einstein's great contribution to the understanding of how classical physics changed for objects when their relative speed approached that of light and led to a symmetry or equivalence between mass and energy — the famous Einstein relation.

Recent advances in physics may be characterized by discoveries that either complete or break some deeply held mathematical symmetry. Breaking a symmetry can have a revolutionary effect on established laws, making it necessary to put more terms in the equations which express the laws. For example, "up" and "sideways" might be symmetric were it not for the introduction of a term in the equations of motion for the gravitational attraction of the earth.

Conversely, the existence of different particle masses must imply a mathematical description which has more terms in its equations than the Standard Model and some symmetry of the multidimensional space must be broken to generate these terms. The symmetry that is to be broken is rather obscure: it is that which constrains the residual potential energy of empty space not to be zero (and hence symmetric) but one sided (in the direction of positive energy).

One of the consequences of the new terms in the theory that stem from this asymmetry is that there must be a heavy particle to be discovered: the Higgs. Clearly we must find it if we are to extend our standard model to include particle masses.

Unless the Higgs Boson has been missed in the vast ocean of data accumulated in the search for the W, Z and the other pieces of the puzzle, it can only be produced at energies out of reach of existing colliders.

In order to have a good chance of finding the Higgs one has to aim for an order of magnitude higher energy than our largest hadron collider, Fermilab's Tevatron, which collides beams of an energy of 1 TeV on 1 TeV. Early plans to build the Superconducting Super Collider in Texas with 20 times the Tevatron energy foundered (see sidebar on SSC) leaving Europe's modest hadron collider of 7 TeV per beam the only machine still adequate for a Higgs Search. This Large Hadron Collider, or LHC, is about to be commissioned.

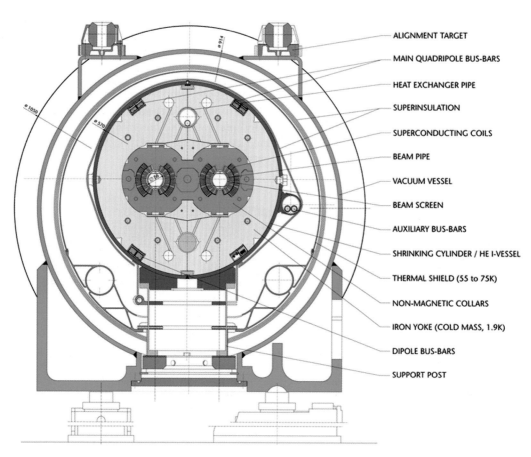

CERN AC/DI/MM - HE107 - 30 04 1999

Fig. 6.15 A picture showing the cross section of the superconducting magnets for the Large Hadron Collider (LHC), scheduled for completion in 2007 at CERN. LHC collides protons on protons and therefore has two beam pipes.

Lyn Evans (1945–)

Welsh accelerator physicist
Project leader for CERN's LHC
Commander of the British Empire (CBE)

Lyn Evans is of a generation of accelerator builders whose career spans the era of colliders — particularly hadron colliders. His research activities started in 1968 with a PhD at the University of Wales, Swansea on the production of plasmas with an intense laser beam. The CO_2 laser, which he built himself, was still in use as late as 1990 in the Swansea Physics Department. This exemplifies a practical approach to the construction and operation of physics equipment which has been his strength throughout his career.

Like many of his contemporaries he came to CERN, once his PhD thesis was completed, as a Research Fellow. His early work involved the investigation of beam dynamics in a 3 MeV experimental linear accelerator, which was to be the prototype of injector linac for the CERN Proton Synchrotron.

In 1971 he moved to the newly approved 300 GeV Project for the construction of the Super Proton Synchrotron under John Adams. His initial contribution was in the field of beam transport, but when the time came to start up the new machine, he made a crucial contribution to overcoming a serious and puzzling instability due to higher resonant modes in the accelerating cavities. When specialists were mainly engaged in playing intellectual football with the many hypotheses about the origin of the instabilities, he took a pragmatic approach, reducing the cooling water to the cavities, which raised their temperature and shifted the frequency of the instability. He went on to find a means to damp the instability, allowing the SPS to exceed its design goal by more than an order of magnitude. It is typical of the man that when frustrated in this work by a scarcity of available professional welders he borrowed a torch himself and did the job. Like many of his countrymen he has enjoyed rugby football and he is not to be deflected by any obstacle which sheer energy and determination can overcome.

He played a key role in the conversion of the SPS into a storage ring and to its successful conversion into a proton-antiproton collider, the first of its kind. He clarified many of the fundamental limitations to the machine performance, which was later to prove of vital importance to the design of the Large Hadron Collider (LHC). In particular, his theoretical and experimental work on intra-beam scattering was essential in elucidating and curing a mechanism that can rapidly degrade the machine performance.

In the mid 1980s, he worked on the commissioning of the world's first large superconducting storage ring at Fermilab near Chicago, applying his intimate knowledge of beam instabilities to overcoming the influence of hidden non-linear fields in the superconducting magnets at injection, and in pushing back the deleterious influence of one beam upon the other when in collision.

In 1989 he was given the task of setting up a new CERN Division responsible for both the operation of the SPS and for the development of the Large Electron Positron Collider (LEP). He once more used his practical approach to elucidate one of the main performance limitations of LEP, the presence of so-called synchro-betatron resonances, and steered it to its design goal in a few years.

In 1994, he was appointed Project Leader of the LHC project, Europe's flagship scientific instrument for the investigation of the high-energy frontier in particle physics. The LHC is a huge project with a materials budget of more than 3 billion Swiss Francs, and enormous technical challenges. He was responsible for both the machine design and the development of the unique system of superconducting dipole magnets capable of reaching up to 9 tesla, operating in super-fluid helium at 1.9 K. He took this development from a laboratory scale to an enormous industrial production.

In 2001 he was awarded a CBE (Commander of The Most Excellent Order of the British Empire) in recognition of his achievements.

---

rate magnet rings and their cryostats it uses the more ingenious solution of putting both beam pipes in the same magnet and cryostat. The magnet cross section of the 7 TeV hadron collider is shown in Fig. 6.15. The proton beams have the same polarity but go around in opposite directions. Thus, although they share a magnetic yoke, they circulate in separate pipes, guided by separate superconducting coils, oppositely polarized.

At the time of writing, the machine is nearly complete, and one hopes that by the time these words reach the reader, it will be contributing to the hunt for the Higgs. The project is being led by Lyn Evans.

VI.10 Heavy-Ion Colliders

So far we have described colliding simple particles — electrons or protons — but as we said in Chapter III, collisions of heavy ions has been under study for a very long time. The Bevalac in Berkeley was used to study ions over the full range of the periodic table colliding with fixed targets at GeV energies. Later this study was extended at the AGS, in Brookhaven, and then at the SPS in CERN.

The idea of extending these studies to colliders, which would provide much higher center of mass energies, became the ambition of nuclear physicists who normally dealt in energies of MeVs rather than GeVs. The quarks contained within a proton or a neutron are held together by a force mediated by "gluons." In a high energy collision these could all be freed, moving throughout the nucleus, forming what is called a quark-gluon plasma. Such a plasma would be a new and previously unobserved state of matter. It would be similar to conditions in the big bang at the start of the uni-

RHIC

The early 1980s found the ISABELLE Project at Brookhaven in a sorry state. Difficulties with the superconducting magnets had prompted the US Government to switch priorities back to Fermilab, where the Tevatron and antiprotons offered the best bet for the future of high-energy research. Later, in 1987, this was to be compounded as the Superconducting Super Collider received the green light. The SSC, however, had its enemies among the science community: the practitioners of so-called "little science." In 1983 Alvin Trivelpiece became Director of DOE's Office of Energy Research and devised a plan to satisfy the SSC's opponents by persuading then-President Reagan to allocate extra money to accelerators within the DOE laboratories to satisfy their needs. These new projects were to be built while the SSC was under construction.

BNL's prize in this game was a new collider: the Relativistic Heavy Ion Collider (RHIC). Heavy ions such as gold would be accelerated in counter-rotating beams and collide, simulating the early history of the universe. In the intense heat generated by the collision, the quarks and gluons, seething within the protons and neutrons of the colliding atoms, would be liberated to form a new and fascinating state of matter — the quark-gluon plasma. The new collider would be nothing other than a modified version of ISABELLE. The new collider was based on the tunnel designed for ISABELLE and the magnets a development of those originally under development for ISABELLE but incorporating advances made at Fermilab and at DESY for their superconducting rings.

The ions would be injected into the collider from Brookhaven's existing Alternating Gradient Synchrotron (AGS), itself fed by its booster and a pair of tandem Van de Graaff accelerators. Naturally these existing machines needed to be adapted to the different charge to mass ratio of the ions. This changes the speed at which the beams circulate and thus requires new radio-frequency cavities whose frequency can be tuned.

The energy of the ions accelerated: 100 GeV per nucleon, sounds rather modest until we remember that a gold ion contains 197 nucleons and that the magnetic constraining forces are provided by only 79 protons. This makes the beam two and a half times harder to bend. The total energy of the ion is 1970 GeV: the highest energy ever given to an atomic nucleus!

Construction started in 1991 and although the engineering of the machine proceeded in a straightforward way it was delayed somewhat when, in 1997, the lab was accused of covering up earlier incidents when tritium had escaped into the nearby landscape. Neither the escape nor the cover-up were true. Even as the start-up approached there were press scares that the high temperatures of the ion collisions and the new quark-gluon plasma might spark off a cosmic explosion. All this had to be answered calmly and scientifically, and the general public convinced.

The collider began operation in 2000 and, as we know, the universe did not explode. Five years later the four detector groups conducting research at RHIC now claim they've created a new state of hot dense matter out of the quarks and gluons, but it is a state quite different and even more remarkable than had been predicted. Instead of behaving like a gas of free quarks and gluons, as was expected, the matter created in RHIC's heavy ion collisions appears to be more like a *liquid* in which all particles have the same velocity.

Satoshi Ozaki (1929–)

Japanese–American physicist
Fellow of the American Physical Society

Born in Osaka, Japan, Ozaki received his BSc in 1953, an MSc in 1955 from Osaka University, and a PhD in 1959 from MIT. He then went to Brookhaven National Laboratory where he remained for the next 22 years. There he engaged in particle physics research. His accomplishments at BNL include particle physics experiments at the Cosmotron and AGS, design and exploitation of beamline and detector facilities, management of the Multiparticle Spectrometer Project (MPS) and the On Line Data Facility (OLDF).

In 1981 he left his position as Group Leader of the Spectrometer Group for Tsukuba, Japan to lead the construction of TRISTAN, a 30 GeV electron-positron Collider at the National Laboratory for High Energy Physics (KEK). He served as the Director of the Physics Department and then as the Director of the TRISTAN Project. Upon completion of the project on schedule in 1987, he became the Director of the Accelerator Department at KEK.

Ozaki returned to BNL in 1989 as the RHIC Project Director. In this capacity, he headed the nine-year construction project to build the Relativistic Heavy Ion Collider (RHIC), a major superconducting collider, and its complement of detectors for high-energy heavy-ion collision physics.

He achieved the goal of completing the RHIC facility, both the collider and detectors, on schedule, and within the project budget (almost 650 M$). RHIC has supplied experimentalists with a number of different colliding ions and thus allowed the exploration of nuclear research in the regime where a quark-gluon plasma is expected to occur.

Ozaki has served on many national and international committees. His advice is widely sought. For example, he served as a member of the Advisory Council for Science Policy and Management of KEK, the SLAC Science Policy Committee, the URA Visiting Committee for Fermilab, the LHC Machine Advisory Committee and the CERN Science Policy Committee.

With RHIC construction completed he is involved in a number of future accelerator projects, like the ILC and NSLS II at BNL.

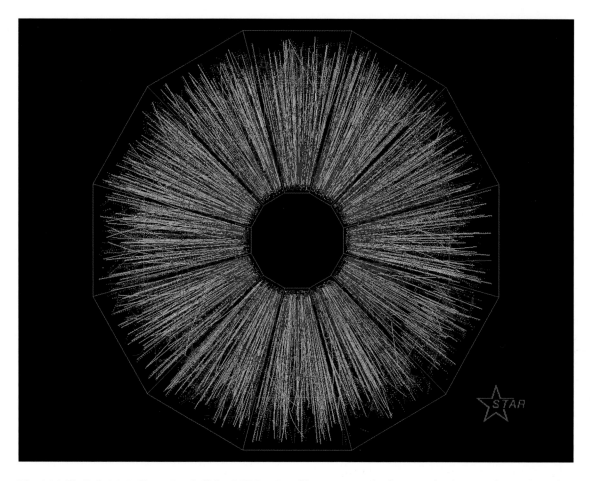

Fig. 6.16 The Relativistic Heavy Ion Collider, RHIC, at Brookhaven National Laboratory, has been used to study nuclear matter under extreme conditions of very high density and very high temperature similar to the conditions in the original Big Bang. Here we see the result of a collision of a nucleus of gold with a nucleus of gold The temperature, in a collision, rises to 2 trillion degrees Kelvin and as many as 10,000 particles are born in the resulting fireball.

verse. It may be created by colliding ions: gold on gold (for example).

In order to do this, the collider RHIC was built at the Brookhaven National Laboratory in the abandoned tunnel of ISABELLE. The new machine — at 100 GeV per nucleon per beam — is superconducting and a very impressive device, built under the guidance of Satoshi Ozaki (see sidebar for Ozaki). Recent results have shown that the quark-gluon plasma is maybe more like a liquid than a gas. The collisions release hundreds of secondary particles (see Fig. 6.16). Whether or not a quark-gluon plasma has been seen is still unclear. There is evidence from a number of different considerations, but as yet no clear "smoking gun" has been seen to confirm it has been produced.

Plans are underway to have heavy-ion collisions in the LHC, and thus to extend the energy available in collisions even further into the TeV per nucleon regime.

Chapter VII. Detectors

VII.1 Early Primitive Detectors

Most particle physics experiments, even today, involve tracing the paths of the particles that emerge from the impact of one accelerated particle with a target. The target may be either a particle at rest in a solid piece of metal or a particle in a colliding beam approaching head on. The particles are themselves too small to be seen or photographed, but they leave a trail of ionized atoms in their wake, which may be detected and recorded. In the very early days when radioactivity was being studied as a new phenomenon, the only means to detect the passage of a charged particle was to peer down a microscope to catch the very faint flash of light produced in a crystal of zinc sulfide or calcium fluoride.

Rutherford had to use this method for his experiments to determine the structure of the nucleus by tracking the particles scattered off it. As early as 1901 C.T.R. Wilson had constructed an electrometer that was sensitive to the ionization of dry air inside a glass bulb, but it was only in 1930, when Geiger became a collaborator of Rutherford, that his primitive fluorescent detectors were replaced by Geiger-Muller counters, that generated an electrical signal and audible pulse.

The counter was a glass cylinder filled with air or an inert gas. High voltage was applied to a very fine wire along the axis of the cylinder. The electric field around such a fine wire can be very much higher than it would be between two flat plates — sufficient to provoke spark or avalanche discharge when a particle ionizes the gas close to the wire. The sudden dip in the voltage supply to the wire caused by the spark produces a pulse, which can be detected by a simple electrical discriminator and amplified, either to make an audible sound or to register in a pulse-counting circuit. Such methods were used in the 1930s for experiments with beams from electrostatic machines and cyclotrons. However in 1944 a novel device, the photomultiplier, began to be used to amplify the faint signals from a fluorescent crystal or plastic scintillator material — the successor to Rutherford's sodium iodide crystal.

VII.2 Scintillators, Photomultipliers and Cerenkov Counters

The development of plastics in the 1940s led to the discovery of a fluorescent transparent material — scintillators — that could be formed into intricate shapes and grafted onto guides for the light. As the particle passes through this "scintillator," it excites electrons surrounding an atom. which emit quanta of electromagnetic radiation as they decay into their ground state. With suitable doping of the material the frequency can be shifted into the visible region.

The invention of the photomultiplier, sensitive enough to produce an electrical pulse from only a few quanta of light, was crucial to exploiting these new materials.

Converting the flash of light into an electrical pulse when coupled with advances in electronic circuits for counting the pulses — first with vacuum-tube diodes, then transistors and finally microchips — hastens the accumulation of data. Logical circuits could be used to detect "coincidences" when signals came simultaneously from two counters, proving that a particle had followed a path through both counters and did not come accidentally from the background of cosmic rays. Placing pairs of counters at the entrance and at a particular exit angle from a magnet could define the momentum of the particle. Nearly all such experiments involve the collection of a very large number of "events," each with several tracks. These are analyzed statistically with an accuracy that improves as the square root of their number. A typical photomultiplier pulse may be as short as a nanosecond, allowing data collection rates of more than a megahertz.

Fig. 7.1 The "Crystal Ball" detector that was first used at SLAC and then at DESY. Here we see a modern array of photomultipiers viewing a nearly-complete sphere of scintillating tellurium-doped sodium iodide crystals.

In Fig. 7.1 we see a modern array of photomultipliers that view a nearly-complete sphere of scintillating crystals: NaI(Tl). A limitation of scintillators is that they themselves cannot establish the identity of the particle. The best one may hope for is to determine the momentum (the product of mass and velocity) with the help of a magnetic field. However, particles of different mass but the same momentum can be distinguished by measuring their velocity. At low energy, the time of flight between two scintillators can be used to find the velocity; at higher energies, the Cerenkov counter does the job.

We will see in Chapter VIII that particles passing through a vacuum produce an electromagnetic wave just like the electrons oscillating up and down a radio transmitter's aerial. A small amount of gas will lower the local velocity with which light travels: in fact, there is an inverse relation between speed and density (or refractive index) of the gas. If the refractive index is large enough, a particle may well be traveling faster than the local velocity of light and produce a shock wave in the form of a cone of light, its apex traveling with the particle like a sonic boom. In a Cerenkov counter this radiation appears as visible light, which can be col-lected by a mirror and focused into a ring. If the beam of particles contains electrons traveling faster than the local velocity of light and also heavier particles traveling slower, only the electrons will produce a ring of light to trigger a photomultiplier. Such a device is called a threshold detector. In more sophisticated Cerenkov counters the diameter of the ring is used to identify the particle, since it is linked to the particle velocity.

VII.3 Collisions in Three Dimensions

Physicists would always like to record the event or collision of one particle with another in three dimensions so that they may visualise the incoming tracks of the particles and those left by the products of the inter-action, fitting them together to explain what happened at the point of collision. Even a complex array of scintillation counters, together with Cerenkov counters, can hardly give a complete picture. They indicate a particle's passage through the instrument but do not indicate its precise path. There would have to be very many of them to distinguish among all possible directions of emerging fragments. They are useful in situations where the particles follow predictable paths, but when we are searching for rare new phenomena another, completely different technique is needed — that of the track chamber.

The first track detectors, which recorded the paths of charged fragments emerging in all directions, were solid blocks or stacks of photographic emulsion. In an emulsion a particle leaves a trail of charged atoms (ions) in its path as it knocks off some of the electrons from nearby atoms. These ions induce exactly the same chemical reaction in the emulsion, as exposure to light. The photographic image can be developed and fixed, and slices of the emulsion viewed and analysed through a microscope. An example is shown in Fig. 7.2. In the earlier days of particle physics research teams with no access to particle accelerators often carried emulsion stacks to the top of mountains or to high altitude in balloons in order to observe cosmic rays.

The ability to repetitively record complete events came with the invention of the cloud chamber. This was a glass sided container filled with air or nitrogen saturated with water vapour. When the piston at one end of the container is withdrawn, the gas cools and becomes supersaturated — ready to condense into droplets at the first excuse. Droplets then form on the recent trails

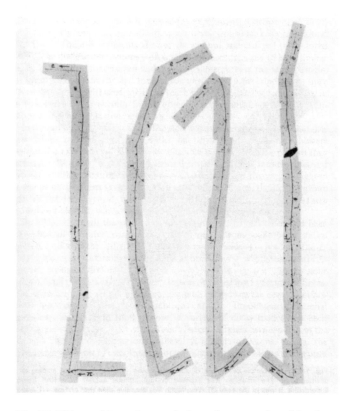

Fig. 7.2 This emulsion micrograph shows four examples of the decay of a cosmic ray pion into a muon and neutrino. In each case the long track is that of the muon, which has a fixed kinetic energy of 4.1 MeV — typical of a two-body decay. At the (upper) end of the long track the muon decays into an electron and two more neutrinos — a three body decay producing different energies of the electron, seen as tracks of different lengths.

Pavel Alekseyevich Cerenkov (1904–1990)

Russian physicist
Discoverer of the Cerenkov Effect
Awarded State Prizes in 1946 and 1951
Nobel Prize, 1958 (with I.M. Frank and I.Y. Tamm)

Pavel Alekseyevich Cerenkov was born in the Voronezh region of Russia in 1904. His parents were peasants. He graduated from Voronezh State University in 1928, and in 1930 he took a post as senior scientific officer in the Lebedev Institute of Physics in the USSR Academy of Sciences. In 1940 he was awarded the degree of Doctor in Physico-Mathematical Sciences, in 1953 was confirmed as Professor of Experimental Physics and, after 1959, controlled the photo-meson processes laboratory.

It was in 1934, whilst he was working under S.I. Vavilov, that Cerenkov observed the emission of blue light from a bottle of water subjected to radioactive bombardment. This "Cerenkov effect," associated with charged atomic particles moving at velocities higher than the speed of light in that medium, proved to be of great importance in subsequent experimental work in nuclear physics and for the study of cosmic rays. Any transparent material — gas, plastic or glass — slows down the local velocity of light wave so that a faster particle will produce a conical wave front like the sonic boom of a jet fighter. The cone can be optically imaged as a ring of light whose diameter is proportional to the angle of the cone and to the excess particle velocity. An even simpler version, the threshold counter, consists of a cylinder filled with gas. The pressure of the gas may be adjusted so that the local velocity of light lies between the velocities of two species of particles that have the same momentum but different mass. Only the lighter particles will then register as a flash of light. The Cerenkov detector was installed in Sputnik III and has become a standard piece of equipment in atomic research for observing the existence and velocity of high-speed particles.

Pavel Cerenkov also shared in the work of development and construction of electron accelerators and in investigations of photo-nuclear and photo-meson reactions. In 1930 he married Marya Putintseva, daughter of A.M. Putintsev, a professor of Russian literature. They had a son, Aleksei, and a daughter, Elena.

of ionised gas left by charged particles. C.T.R. Wilson, whose aim was, surprisingly, to reproduce the atmospheric conditions of a Scottish mountaintop, invented this device as early as 1912. Anderson used such a device to discover the first particle of antimatter, the positron, in 1932. We see an example in Fig. 7.3.

Much later, in 1952, a similar idea was applied by Glaser to supercooled liquid hydrogen in which the ionized trails of particles appeared as a trail of bubbles,

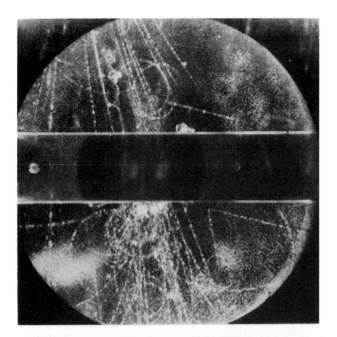

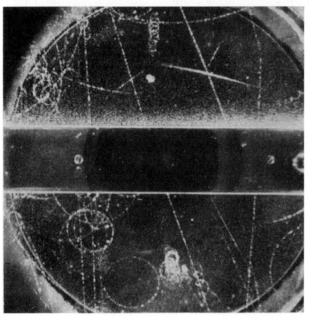

Fig. 7.3 These are two cloud chamber pictures from 1947.

Donald Glaser (1926–)

American physicist
Awards from the Physical Society of London and the American Physical Society
Nobel Prize, 1960
Member of the National Academy of Sciences

Born in Cleveland, Ohio and educated in elementary and secondary schools there, he studied the violin and viola, becoming a member of the Cleveland Philharmonic Orchestra at age sixteen. His undergraduate work was at the Case Institute of Technology, and his graduate work was at the California Institute of Technology, where he received a PhD in 1950. He then went to the University of Michigan where he was made a full professor, at age 31. He invented the bubble chamber while at Michigan, but left there in 1959 for the University of California, Berkeley, where he has remained ever since.

After receiving the Nobel Prize he switched his interest to biology. His development of automatic devices for preparing, measuring, and the handling of large data samples is now put to good use every day in biology. His work included studying regulation of cell growth and the causes of mutation. Currently he uses psychophysical and theoretical methods to study the human visual system, especially mechanisms for sensing motion and the role of cortical noise in visual perception.

Glaser has been able to keep up his interest in music by playing with chamber music groups. In addition he has had a lifelong enjoyment of mountain climbing, skiing, sailing, and kayaking.

which could be photographed. An example is seen in Fig. 7.4. Large bubble chambers were constructed, several meters in diameter and often surrounded by powerful magnets. The momentum of the particles could be calculated from the curvature of the track in the magnetic field. The stronger the field and longer the track, the higher the precision.

Analysing photographs of these tracks became quite an industry at the big accelerator labs in the US and at CERN in the 1960s and 70s. They were projected onto tables in a darkened room where small armies of "scanning girls" used optical devices linked to early computers to convert several points on each track into numbers stored for later analysis by the computer. Sophisticated pattern recognition programs were developed to identify the few interesting events in hundreds of thousands of photographs. During this period a surprising number of marriages ensued between scanning girls and male physicists who had been directing their work.

Bubble chambers could record the tracks of a few particles emerging from a collision every few seconds. Like the synchrotrons that produced the projectile particles, they took a couple of seconds to recover and prepare themselves for another "event." However, collider rings like the ISR produced a continuous stream of events and there was an urgent need to speed up the tracking process and record tracks in an electronic form that could be fed directly into a computer for analysis and visual reconstruction. We see a modern example of this in Fig. 7.5. Particle detectors that produced an electrical signal — Geiger-Muller and proportional chambers — had been long used to monitor radiation levels but it was not easy to apply these to three-dimensional problems.

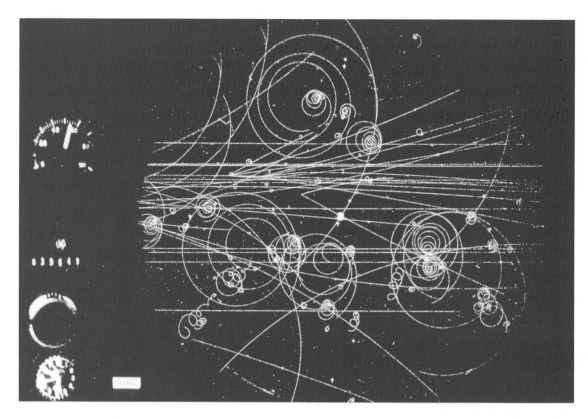

Fig. 7.4 Photograph as 16 GeV negative pions (π-mesons) traverse the first CERN liquid hydrogen bubble chamber; it was only 30 cm in diameter.

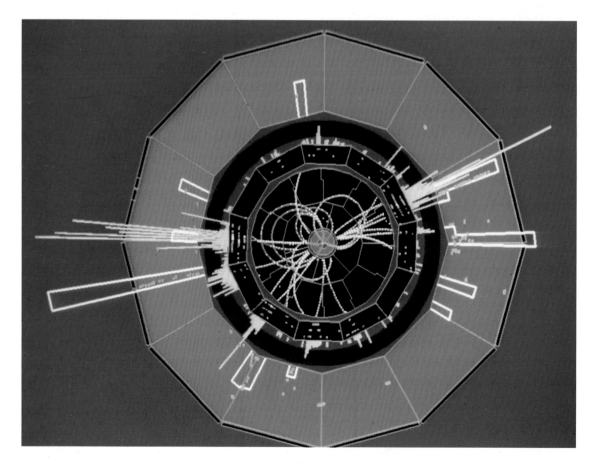

Fig. 7.5 This is a computer reconstruction of decay of a Z^0 in one of LEP's detectors — ALEPH. By this era, in the 1980s, solid state detectors whose output was electrical, and was analyzed by computers, were coming to the fore.

The first way to speed up the data taking was to use a spark chamber. This was a stack of metal plates spaced by a few millimetres. An alternating voltage is applied across neighbouring plates. If the voltage is high enough a charged particle triggers an electrical breakdown between the plates, just as in a Geiger counter and shows the point where the particle crosses the gap. Of course the trail of sparks still had to be photographed and scanned.

Another development was to use more widely spaced transparent electrodes and apply a very short, 10 nanosecond high-voltage pulse to them. The spark appears as a streamer about a millimeter long following the particle's track. This too had to be photographed and scanned.

There was yet another kind of chamber in which the discharge took place between planes of fine wires. The electric field rises rapidly close to a wire and electrons are accelerated to a high enough energy to produce more ionization on impact with the wire. The wire chamber led on to the breakthrough by Charpak in 1968, which finally eliminated the need for human scanners. Charpak's chamber was based on proportional counters — cousins of the Geiger counter. In these counters, electrons are produced as in the Geiger counter, and accelerate as they approach a fine wire anode. They ionize atoms in the filling gas and produce a controlled avalanche, multiplying the charge by a field gain of the order of 10^5. The signal that reaches the wire is proportional to the initial number of ions and strong enough to be picked up by microelectronic circuitry at the end of the wire. The time it takes for the signal to reach the end of the wire gives precise position information in the direction parallel to the wire and the accuracy, less than a millimetre in each direction, is impressive.

Charpak's chambers were later adapted into whole-body PET scanners used in medicine — an unexpected spin-off of the technology of particle physics. These devices record a back-to-back pair of gamma rays pointing to a positron emitted by tiny amounts of radioactive elements attached to biochemical substances. When injected into a patient the positron emitters follow the blood flow and can be inserted into chemicals that home in on tumours. Reconstructing the origin of the radioactive emissions reveals a map of the body with these features highlighted.

VII.4 A Modern Detector

Modern storage ring colliders have a number of points on their circumference where particles collide and

Georges Charpak (1924–)

Born in Dabrovica, Poland
Naturalized French citizen, 1946
Nobel Prize, 1992

Georges Charpak studied at the prestigious Ecole des Mines in Paris where he graduated with a Bachelor of Science degree and went on to complete a PhD for experimental research in Nuclear Physics at Collège de France.

From 1948 to 1959 he was at the Centre National de la Recherche Scientifique (CNRS) and from 1959 to 1991 at CERN where, in 1960, he participated in the first exact measurement of the magnetic momentum of the muon. In 1968 he invented multiwire proportional chambers followed in 1974 by spherical drift chambers for studies of proteins by x-ray diffraction. His next contribution was to multistage avalanche chambers and the application of photon counters for the imaging of ionizing radiations.

In the years from 1985 to 1991 he participated in experiments at Fermilab and introduced luminescent avalanche chambers. At the same time he developed instrumentation for biological research using beta-ray imaging at the Centre Médical of the University of Geneva.

In 1992 he was awarded the Nobel Prize in physics for his invention and development of particle detectors, in particular the multiwire proportional chamber.

The announcement explained that sometimes only one particle interaction in a billion is the one searched for. The experimental difficulty lies in choosing the very few exceptionally interesting, particle interactions out of the many observed. Photographic methods, once so very successful in exploring particle processes, are not good enough for this. In the new wire chamber Charpak used modern electronics and realized the importance of connecting the detector directly to a computer. The invention made it possible to increase the data collection speed for registering charged particle trajectories by a factor of a thousand compared to previous methods. At the same time the spatial resolution was very often considerably improved. His fundamental idea has since been developed and applied for more than three decades. Charpak continues at the forefront of this development. The development of detectors very often goes hand in hand with progress in fundamental research. Various types of particle detectors based on Charpak's original invention have been of decisive importance for many discoveries in particle physics. Several of these have been awarded the Nobel Prize in physics. Charpak has actively contributed to the use of this new type of detector in various applications, for example, in medicine and biology.

As a survivor of the notorious Dachau concentration camp in World War II, Charpak has been an ardent defender of human rights throughout his career.

where there are huge cylindrical detectors that represent another level of sophistication in experimental technique (see Fig. 7.6). These surround each collision point with a series of concentric cylinders so that almost all the particles emerging are recorded in the cylinder or its "end caps." The function of the detector is to reconstruct the tracks of particles emerging from the collision, identify the particles and establish their charge and energy using one or more of the different techniques described below. All this information is piped through fast electronic circuits to a bank of computers where it is recorded and analyzed off-line to search for rare particles or to study the laws of physics that govern the weak, strong and electromagnetic forces between the particles.

The innermost concentric cylindrical shell is usually a "vertex" detector placed as close as possible to the collision point. A typical vertex detector is an array of layers of postage-stamp-sized electronic chips surrounding the beam pipe on concentric cylinders. Each chip; a single piece of silicon, is like an electronic checkerboard with about 40,000 squares on one side and its own readout and is similar in form to the recording elements in many video cameras.

Particles traveling through the thin chips leave behind small electric charges in the squares they cross.

The location of these charges can be recorded electronically and a computer can reconstruct the tracks of all the particles through the layers. Since the electronic squares are so small, they allow measurement of the charged particle position with microscopic accuracy of about one micron.

Charged particles are never created alone, but always in pairs of equal and opposite charges and by reconstructing each path back to where it meets with one or more other paths we can find the position — the vertex — where any given charged particle was created. One of the purposes of measuring particle tracks very close to the interaction point is to identify those tracks that do not come from the vertex and which could be, for example, a signature of short-lived decaying particles. Precise drift chambers — descendants of the wire chambers invented by Charpak — have also been used as vertex detectors.

The next layer around the vertex detector is normally a particle "tracker." In these devices the momentum of a charged particle is inferred from the bending of its trajectory when passing through a strong solenoidal magnetic field. In low luminosity colliders a Time Projection Chamber (TPC) is often used. This kind of chamber was first invented by David Nygren of Berkeley back in the 1970s for use on PEP at SLAC.

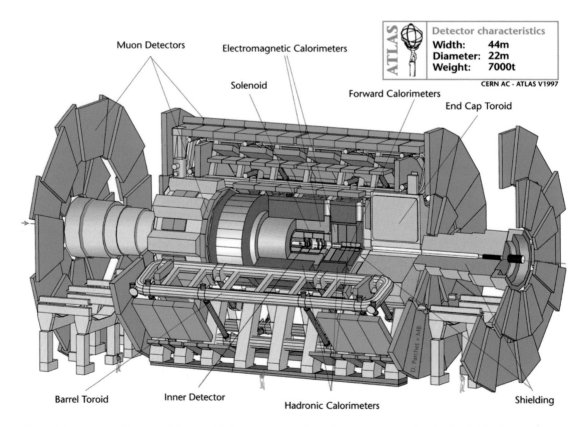

Fig. 7.6 A cut away diagram of the ATLAS detector, presently under construction, for the LHC. The huge scale can be seen from the tiny figures.

A TPC is a cylindrical chamber, filled with pressurized gas in a perfectly uniform electric field. The chamber is bathed in the magnetic field of a huge solenoid and the curvature of the charged tracks in this field gives their charge and momentum. As particles pass through the chamber they collide with atoms of gas, knocking off a string of electrons that respond to the electric field and drift towards the end of the chamber. At the end of the TPC is a pie-shaped array of wires and sophisticated electronics that detect and sift through thousands of particles simultaneously and can clearly distinguish between similar types of particles (see Fig. 7.7). One of the inherent limitations of the TPC is that the electrons drift relatively slowly in the electric field (some mm per μsec) making it very difficult to disentangle tracks arising from different events. For this reason, detectors at high luminosity colliders, such as the LHC, use either silicon based trackers or gas-filled devices in which the drift distance to the detecting wire is of the order of a few millimeters.

Short-lived neutral particles may be identified among the tracks recorded in the vertex chamber when charged particles are seen to originate "out of thin air" just a few millimeters from the vertex. Unambiguous identification is more complicated. Each particle coming from the vertex produces a shower of particles which are stopped in the dense outer cylindrical shells of the detector called "calorimeters." They typically are an alternating sandwich of absorber and scintillator designed to measure the total energy of the shower and hence deduce the energy of the parent particle.

Earlier we mentioned that a Cerenkov counter measures the velocity of a particle and, coupled with a measurement of momentum, identifies the mass of the particle. In the same way a measurement of energy in the calorimeter, together with momentum determination, can tell us the mass and identity of a particle. Different calorimeters can be used to distinguish between electromagnetic particles, which convert (gammas) or stop (electrons or positrons) in the inner shells, and the much more penetrating hadrons, stopping in the outer layers.

The only missing charged particles, which survive to register in the outermost layer of counters, are muons. Collider experiments are covered in layers of chambers that detect these muons and measure their momentum from their curvature in the magnetic field. Any particles that escape detection are neutrinos (which are uncharged) whose presence can often only be deduced by identifying all the other particles and looking at the known rules for decay that can point to an accompanying neutrino.

In sum, a modern detector is a huge and very complicated thing. (See Fig. 7.8.)

VII.5 Digital X-ray Imaging

The techniques developed for a digital imaging of the particles at the vertex of a modern collider experiment have produced spin-offs for the fields of medical diagnosis and security.

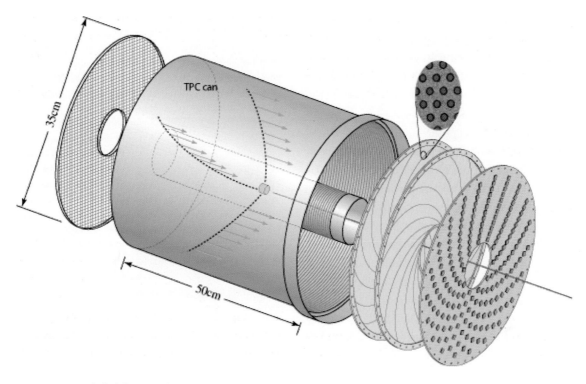

Fig. 7.7 An exploded diagram showing a Time Projection Chamber (TPC). A description of this complicated detector is given in the text.

Fig. 7.8 A picture of a modern detector; in this case the central solenoid of the mammoth CMS detector at CERN. One can easily imagine that these detectors are as complicated as the accelerator itself.

Photographic film, the most widely used detection medium in x-ray imaging applications, has been generally used since the discovery of x-rays at the end of the nineteenth century. However, photographic emulsion absorbs only a tiny fraction of the energy of x-rays passing though it. A more substantial recording medium, such as flat array of semiconductors similar to that used for a vertex detector, would be better.

Such x-ray imaging systems with their digital readout make the image immediately available on a monitor. This not only saves processing time but allows on-line aids to help the user analyze the image. For example, the image can be colored to identify interesting areas more easily, and comparisons may be made by subtraction. Perhaps the most important aspect is that the new imaging techniques have greatly reduced the radiation dose of medical x-rays. Thus the risk of inadvertently causing cancers with diagnostic x-rays is now negligibly small.

The most common modern sensor concept is the silicon based CCD (charge coupled device). Detector materials other than silicon such as gallium arsenide, cadmium telluride, and cadmium zinc telluride, are also under extensive investigation. The advantage of all these materials is their higher absorption coefficient resulting in a higher sensitivity compared to photographic film. The higher performance may be used to obtain images with higher signal to noise ratio or to reduce the radiation dose.

Detector arrays for digital x-ray imaging systems may be of either the linear or area type. The linear type consists of one or a few rows of detector pixels, and requires that the object of interest be scanned over the detector. This is ideal, for example, for inspection at airports where luggage passes on a conveyor belt. On the other hand, area arrays are two-dimensional, filled with rows and columns of pixels; require no scanning procedure. Such arrays have revolutionized the use of x-rays on-line in the operating theater, particularly in heart surgery where restricted blood flow can be relieved with fine catheters inserted at more convenient parts of the arterial system. Patients (and doctors) no longer have to fear hazardous doses of radiation during these procedures.

A typical flat panel sensor, 40 cm square, is fabricated using thin film technology based on amorphous silicon technology. The incident x-rays are converted by scintillator material into visible light which generates electron hole pairs in sensors, which are similar to conventional photodiode arrays. Each pixel in the array consists of a light-sensing photodiode and a switching thin film transistor (TFT) in the same electronic circuit. Amorphous silicon photodiodes are sensitive to visible light, with a response curve roughly comparable to human vision. The sensitivity of amorphous silicon photodiodes peaks in green wavelengths, well matched to scintillators. The charge carriers are stored in the capacitance of the photodiode. By pulsing the gates of a TFT line within the matrix, the charges of all columns are transferred in parallel to the signal outputs. All signals of the columns are amplified in custom readout multiplexers for further processing.

The same technology can be put to good use in the field of security allowing on-line, real-time scanning of the contents of luggage, freight containers and trucks as they pass a linear array. It is possible to scan the whole truck, from driver cabin to wheels, in this way. There are sophisticated techniques which allow differentiation between organic products such as drugs or explosives, and inorganic material, such as steel, to allow easy identification of grenades, knives, and other weapons.

VII.6 Detection Techniques for Synchrotron Radiation Sources

We shall see in the next chapter that there is a large community of scientists who use synchrotron radiation for their research. This radiation is emitted tangentially from a high energy electron beam as it circulates in a synchrotron. The radiation, like white light, which contains a continuous spectrum of wavelengths, also covers a broad spectrum, but the wavelengths are much shorter, in the ultraviolet and x-ray range. We can think of the beam of synchrotron light as consisting of light "particles" or photons of a vast range of energies. Users of synchrotron radiation sources, like their counterparts in nuclear physics, study scattering. Although they are not interested in tracking individual photons, the interference pattern produced when the photon beam is scattered reveals the structure of the object under investigation — often a complex molecule or crystal structure. The scale of the detail investigated determines the wavelength of the synchrotron radiation used. In most cases it is in the extreme ultraviolet and x-ray spectrum.

Photographic films, identical to those used for the medical applications of x-rays, could be employed to record the scattered pattern, but modern machines all use electronic recording of x-rays, using pixels which are sensitive to radiation that convert the incident photon into an electric signal (like digital cameras and camcorders).

As we remarked, the spectrum of radiation is continuous. Most experiments, however, require just a very small band of wavelengths (an almost monochromatic beam); only then will the interference patterns from the scattering be clearly resolved and not overlaid by different patterns from other wavelengths. The small slice of wavelengths for an experiment is selected with filters, called spectrometers or monochromators; vital to any synchrotron radiation facility. An example of a simple spectrometer in the visible range is a prism or diffraction grating that breaks up white light into a band of colors, an effect shown many hundreds of years ago. The equivalent to a diffraction grating in the x-ray range is a crystal.

The regular structure of a crystal forms a series of interference fringes where the electromagnetic waves of light reinforce or cancel depending upon the angle they emerge — again, just like a diffraction grating. The interval between fringes depends upon both the wavelength of the radiation and the spacing of the atoms in the crystal. A crystal at a glancing angle will send x-rays of only a very specific wavelength in a particular direction and so form a monochromatic beam.

With such a beam, together with a device to detect and record the x-rays scattered from a sample, an experimenter can take the three-dimensional diffraction patterns of rings and use a computer to deduce the original shape of the crystal or molecule. In the following chapter we discuss how synchrotron radiation is produced.

Chapter VIII. Synchrotron Radiation Sources

VIII.1 Scientific Motivation

Elsewhere in this book we have described how large synchrotrons and colliders are used for particle physics research. We have also explained how smaller accelerators are used extensively for practical applications including a variety of industrial processes and, in medicine, to diagnose and cure patients of cancer. In the next chapter we shall discuss the medical use of accelerators. This chapter is devoted to the synchrotron radiation source: a class of electron accelerator which, though more modest in size than its siblings in particle physics, is also built to advance our knowledge of the world around us. The beams of synchrotron radiation — ultraviolet light and x-rays — that they produce are ideal for a spectrum of studies, from crystallography to enzyme biology, with great implications for both today's science and technology and tomorrow's.

Lightweight particles such as electrons radiate "synchrotron radiation" when traveling very close to the velocity of light in a regime that can only be explained by Einstein's special theory of relativity. The radiation emerges in an intense and narrow cone which is tangential to their circular path as they are bent by the magnets of the synchrotron. A very broad spectrum of frequencies is emitted simultaneously. It can extend from the long wavelengths of red light through blue and ultraviolet into the x-ray range. The peak of the spectrum may be shifted to shorter wavelength by bending higher-energy electrons in stronger magnetic fields.

The huge range of wavelengths makes synchrotron light a powerful tool for unraveling the structure of materials, crystals and molecules from the microscopic to the atomic scale. The wavelength can be short compared to the detail in the structure investigated. It is the x-ray part of the spectrum that is particularly important in research, as the intensity of this radiation is orders of magnitude greater than can be obtained from any other source.

The x-rays from synchrotrons are being used by more than 20,000 researchers at 54 dedicated facilities in 19 different countries. There are eight more facilities under construction, and eleven more in the design and planning stage. Through the years, the intensity of x-ray beams has grown. Between 1960 and 2000 it first increased by a factor of a billion and then by a further factor of a million more. In fact, this rate of growth was even more than that of computer performance, which is otherwise considered the most rapid the world has ever seen.

Studies with synchrotron radiation range over much of science. There have been 18 Nobel Prizes awarded for work related to x-rays: 8 in chemistry, 7 in physics and 3 in medicine. Surely there will be more.

In physics, the work with x-rays from synchrotron radiation elucidates the structure of matter, the magnetic properties of the material for hard discs for storing digital information, and the origins of high temperature superconductivity. Surface phenomena studies include the wear of aircraft turbines as well as the behavior of catalysts for industrial chemical processes.

In biology and medicine there have been studies of the structure of DNA and of protein, studies of osteoporosis, of diseased human hearts, and studies of the fatal activity of anthrax and cholera toxins as they attack a cell. Of particular importance is the use of synchrotron radiation in the design of drugs, and in the configuration of enzymes for various industrial processes. A picture of a protein strand showing the detail possible with synchrotron radiation is seen in Fig. 8.1.

In environmental science and geosciences there have been studies of the motion of toxic and radioactive wastes to address the danger of contamination of water supplies and the design of appropriate filters for removing toxic substances from run-off.

In art and archaeology: x-rays from synchrotron light sources as well as from more modest sources have

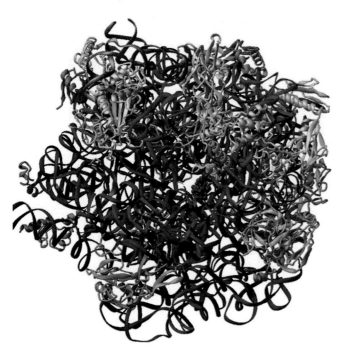

Fig. 8.1 This intricate structure of a complex protein molecule structure has been determined by reconstructing scattered synchrotron radiation.

helped in the restoration and identification of art works. They were used to determine that it was sulfuric acid that was causing the decay of the Vasa, a ship sunk in the Stockholm harbor in 1628. This treasure has been brought up, fully restored, and is now a popular tourist attraction.

There has been growing interest in the archaeological and historical use of synchrotron radiation. For example, at the synchrotron radiation laboratory in Daresbury, England, 5% of the machine time is devoted to this purpose. In Trieste, the synchrotron facility, Elettra, is used to study the composition of ink found in Pompeii in almost perfect condition. Learning the ink's make up will help in restoring Roman manuscripts. Facilities in Australia, Canada and the Middle East all have declared cultural heritage as an integral part of their programs.

VIII.2 Principles and Early History

The first step in understanding synchrotron radiation was taken in 1865 when James Clerk Maxwell produced a set of equations that unified electricity and magnetism and, amongst other things, predicted electromagnetic radiation. Refined versions of these equations, by one of the giants of physics in the 19th century, are still in use today. They are used to describe all classical electromagnetic phenomena, including the emission of synchrotron light.

The next step was in 1887 when Heinrich Hertz demonstrated that — as predicted by Maxwell's equations — changing charge densities and changing electric currents produced by a spark gap generated electromagnetic waves that, like visible light, propagated over large distances. That, of course, was the start of radio communication, radar, TV, cell phones, the GPS system and most of all the technical aspects of modern life, for electricity and magnetism are at the heart of almost all modern technological devices.

One of the missing links was the realization that electric currents were really streams of electrons. As early as 1858 "cathode rays" had been discovered; that is, rays that emerged from the negative electrode — cathode — of electric discharge tubes and were accelerated to the positive anode. In 1883, Hertz showed that these rays could be bent by a magnetic field, but it was not until 1897 that J.J. Thompson definitively demonstrated that these rays were made up of electrons. It was he who really discovered the electron, and who determined both its charge (equal to that of a proton, but of opposite sign) and its mass (2000 times smaller than that of a proton).

One might be tempted to think that to understand synchrotron radiation from an electron undergoing acceleration would require all the paraphernalia of modern quantum mechanics and relativity, but the basic idea is very simple. Imagine that you were able to attach an electron to the end of a stick and wave it around. It would send out electromagnetic waves that would be very weak but no different in nature and no more mysterious than the radio waves emitted by electrons flowing up and down in a transmitting aerial. In order to produce a detectable signal one would need many electrons and have to wave the stick very vigorously and preferably close to the velocity of light! An electron coasting along in free space emits no such radiation. It is the force applied to an electron — causing it to slow down, accelerate or be deflected sideways — that produces the radiation. Some readers who have met Maxwell's equations may remember that field is produced by "changing" currents — equivalent to accelerating charges.

The first prediction of synchrotron radiation based on Maxwell's equations appeared in an obscure French journal appropriately called *Lumière et Eclairage*.

Now imagine an intense bunch of electrons being bent sideways by one of the many dipole magnets in an electron synchrotron. The bunch produces radiation as it passes through the magnet. The radiation is not of a single wavelength but rather a broad spectrum of frequencies. The same behavior occurs in sound waves. A pure note is a sine wave going on forever but if the sine wave only lasts a short time it is heard as a spectrum of frequencies and the width of the spectrum increases, as the pulse is made shorter.

We now must bring in Einstein's principle of relativity, which has a very strong effect. According to this, as the electron approaches the velocity of light, the intensity of the radiation increases and it is concentrated in a narrow cone along the path of the beam. This narrow beam shines out on a tangent to the synchrotron ring and may be directed onto a sample, for example a crystal.

In 1908, using these newly developed tools, G.A. Schott was able to present a definitive theoretical study of the radiation from a single electron.

VIII.3 Synchrotron Radiation

The observation of the effect of radiation of an accelerating electron, later to be called synchrotron radiation, did not occur until 1947, when it was detected in a betatron. The betatron had been invented by Donald Kerst in 1941 (as described in Chapter IV), and after World War II a 100 MeV betatron was built at the General Electric Laboratories. This machine had a ceramic vacuum chamber, so radiation did not escape into the outside world, but in 1946 John Blewett observed that the rate of decrease in the radius of circulating electrons was consistent with the electrons radiating away energy. This discovery was soon confirmed when radiation was visually observed on a synchrotron by E. McMillan.

Blewett's outstanding discovery was not given the attention it deserved. Taylor and Hulse were awarded a Nobel Prize for a similar piece of scientific detective work when they were the first to observe gravity wave radiation through its effect in changing of the rate of rotation of binary stars.

In 1945, McMillan and Veksler invented phase focusing (as described in Chapter V) and betatrons were soon abandoned by high-energy physicists in favor of synchrotrons. In 1947, the world's second operating synchrotron (the first was only at 8 MeV) was built at General Electric. This 70 MeV machine had a glass donut, transparent to synchrotron radiation — which to everyone's surprise was in the visible range. At first it had been thought to be in the radio spectrum, but no amount of crawling round the machine brandishing what then served as a portable radio revealed the expected signal. It was only when they took a photo of the machine in operation that they saw a bright beam of light. The whole Board of Directors were brought to see it (through a mirror), and in the next two years there were visits from many scientists, including 6 Nobel prizewinners.

The definitive theory of radiation from arbitrarily moving relativistic electrons was published three years later in 1949 by Julian Schwinger (himself a Nobel laureate for his contribution to the development of quantum electrodynamics). This work, which we rely upon today, is more elegant and complete than that of Schott.

Much of the broad electromagnetic spectrum emitted from circulating electrons in a synchrotron can be made in other, and often simpler ways. To take an example: furnaces, or hot elements of electric fires, are copious sources of infra-red radiation; while lasers have been highly developed to work in the visible, or near-visible, range. However the photons of synchrotron radiation produced in the x-ray range are unique. Synchrotron radiation is far more intense than any other source of x-rays. Even in the simplest early synchrotron sources, the intensity was a million times greater than typical dental or medical x-ray machines.

VIII.4 First Generation Synchrotron Sources

At first, synchrotron radiation was obtained from machines built for other purposes: mainly for the study of high-energy physics. These machines are called First Generation Sources and prior to 1973, about 10 synchrotrons had been used in this way. These included machines in Germany, Denmark, the Soviet Union, Japan, and Sweden. There were also four storage rings, one in France, one in the Soviet Union, and two in the United States. These storage rings had the advantage over synchrotrons in that they provided very stable and continuous x-ray beams.

By far, the most energetic of the first generation sources was the Cambridge Electron Accelerator (CEA) in Massachusetts, but the machine was shut down in 1973, shifting the activity to SLAC, where a new storage ring fed by the two mile linac was available (see sidebar on the CEA). This machine, SPEAR, was perhaps the most effective of the first generation machines and, later, it was upgraded to become a third-generation machine.

VIII.5 Second Generation Synchrotron Sources

In the beginning of this science, the early 1970s, there were just a few scientists using the radiation from these first generation facilities. It was very soon realized that optimized rings could produce radiation with many special and desirable aspects, but this step had to await the development of a user community. However, gradually, the interesting scientific results obtained from the first generation stimulated ever more interest and, as the community of users grew, pressure developed to build a second generation of dedicated facilities.

The distinguishing feature of second generation machines is the use of "wigglers" and "undulators." These devices consist of an alternating sequence of dipole (bending) magnets, which deflect the electron beam, first to the left and then to the right. The magnetic field of the bending magnets in a wiggler is stronger than that employed in bending the beam in the circular path of the synchrotron. Of course the beam wiggles and not the magnets. The curvature of the orbit is greater and the radiation emitted greatly enhanced. They are called insertion devices because they are inserted into the regular pattern of the quadrupoles and bending magnets, which form the lattice of the ring, without too much perturbation to the ideal path of the beam.

The development of these devices was pioneered at the CEA, where they were first employed not to give enhanced synchrotron radiation, but rather to control the motion of the stored electrons.

The first wiggler built to enhance radiation was built in Berkeley in 1978 for use at SPEAR. It proved to be very effective and most of the dedicated second generation machines were built with very long straight sections that would accommodate such insertion devices. Herman Winick was the person largely responsible for promoting these devices. The use of permanent magnets for this purpose was pioneered and developed by Klaus Halbach.

Undulators, the first of which was built in 1980 at SPEAR, are simply wigglers where the deflection is not as severe. The radiation from a number of gentle bending sections can interfere to produce a limited but very intense spectrum, which is useful for many experiments. It is particularly of interest when the sample must not be overheated.

An interesting present day Second Generation project, SESAME, is described in the sidebar. (See sidebar for SESAME as well as Fig. 8.2.)

VIII.6 Third Generation Synchrotron Sources

Given the great success of the large number of second generation facilities, user pressure developed to go to even shorter wavelengths; i.e., to even more energetic ("harder") x-rays. This meant that the facilities had to have even higher energy electrons and thus a larger "footprint." These facilities employed the very latest in machine design and the latest insertion devices to produce very intense x-ray beams. The first of this new generation was the European Synchrotron Radiation Facility built at Grenoble (France), which stores 6 GeV electrons (see Fig. 8.3). It was followed by the Advanced Photon Source (APS) at Argonne, USA with 7 GeV electrons and then SPring-8 (Japan) with 8 GeV electrons (see Fig. 8.4).

At the same time that there was the push to develop second generation sources, described in the last paragraph, synchrotron radiation sources that employed lower energy electrons (in the range of 1–3 GeV) took advantage of the improvements in the larger Third Generation

Herman Winick (1932–)

Fellow of the American Physical Society
Fellow of the American Association for the Advancement of Science
Humboldt Senior Scientist Award, 1986
US Particle Accelerator School Prize for Achievement in Accelerator Physics and Technology, 1995
US Department of Energy Distinguished Associate Award, 2000
New York Academy of Sciences Heinz Pagels Award for Human Rights, 2005

Herman Winick was born in New York City, received his AB degree in 1953, and then his PhD in 1957 from Columbia University. He did post-doctoral work at the University of Rochester during the years 1957–1959 and then moved to the Cambridge Electron Accelerator at Harvard University where he served as Head of Operations and Assistant Director. In 1973 he became a Senior Research Associate, and then a professor, in the Department of Applied Physics, Stanford University. He was Deputy Associate Director of the Stanford Synchrotron Radiation Laboratory from 1973–1996 and has remained associated with the sychrotron radiation facility until and even beyond his retirement in 1998.

For more than 30 years he has played a leading role in the development of synchrotron radiation sources and research at Stanford and around the world. During his career he published more than 100 scientific articles, edited several books an synchrotron radiation science and sources, served on editorial boards for scientific journals, handbooks, and encyclopedias, and served in, and chaired, review and advisory committees for synchrotron radiation laboratories in Armenia, Australia, China, Germany, India, Japan, Taiwan, the UK, and the US. He is particularly well-known for his work on radiation sources, wigglers, undulators free-electron lasers and high-brightness electron sources.

He has a strong interest in human rights and in the 1970s took up the cause of the dissidents Fang Li Zhi, Liu Gang, and others in China. He met with Soviet dissidents on trips to the USSR in the 1980s and served on the Sakharov, Orlov, and Sharansky (SOS) Committee to free these and other Soviet dissidents. He was a member of the APS Committee on International Freedom of Scientists (CIFS) from 1989–93 and chaired this committee in 1992. He is currently Chair-elect of the APS Forum on International Physics.

He has a strong interest in international scientific collaboration, and has been associated with many facilities around the world. He initiated and has taken a leadership role on the SESAME project, a synchrotron radiation facility in the Middle East to be shared by Arab and Israeli scientists, now in construction in Jordan.

He enjoys spending time with family, as well as sports and pastimes that include tennis, swimming, hiking, skiing, and traveling.

Klaus Halbach (1924–2000)

German-American physicist
Arthur H. Compton Prize
US Particle Accelerator School Prize for Achievement in Accelerator Physics and Technology

Born in Wuppertal, Germany, Klaus Halbach received his PhD in nuclear physics from the University of Basel, Switzerland. After a three-year teaching experience at the University of Fribourg in Switzerland, Klaus came to the US in 1957. There he worked at Stanford University, continuing his thesis topic, but now under Nobel Prize-winner Felix Bloch, a pioneer of nuclear magnetic resonance (NMR). Throughout his life, Halbach revered Bloch as both colleague and teacher. Following a short return to Switzerland to start a plasma physics group, he joined the plasma physics group at Berkeley in 1960. His work with plasma physics led him into accelerator design and he was a major contributor to the Omnitron, a synchrotron that would have accelerated nuclei from hydrogen to uranium. Though never built, the Omnitron's design laid the groundwork for future accelerators of similar purpose.

Halbach contributed to accelerators in many different ways, but is best known for his work on magnetic systems for particle accelerators. He and his colleague (and later son-in-law) Ron Holsinger developed a package of computer codes still in use today. Halbach become one of the world's premier designers and developers of permanent magnet systems used primarily in wigglers and undulators, and increasingly in conventional accelerators. His contribution to the development of wigglers and undulators, in synchrotron radiation sources, was crucial to the worldwide development of the so-called third-generation light sources. Halbach was a consultant to many projects around the world, including the Advanced Photon Source at Argonne and the Stanford Synchrotron Radiation Laboratory. All of the principal radiation sources within these machines depend on the permanent magnet technology now known as the Halbach Array. He was also a major contributor to the designs of high-resolution spectrometers at Jülich and LAMPF, Los Alamos.

He made contributions to a diversity of other projects such as magnets for a miniature cyclotron that could be used for medical radioisotope production, magnets and low-friction magnetic bearings for an electromechanical battery, and the design of miniature permanent magnet NMR spectrometers for future Mars lander missions.

To his colleagues and PhD students Halbach was above all an outstanding teacher. The technical lectures he delivered were of transparent clarity and presented in a way that invariably held the attention and interest of his audience. But more important was his ability — one might almost say his compulsion — to impart his ideas to anyone willing to listen. He had a love of physics, and he was able to convey how it really was fun. Klaus's colleague, Brian Kincaid, coined a phrase which might apply to dozens of physicists around the world who had enjoyed his company: "alumni of Halbach University."

SESAME

SESAME is short for "Synchrotron light for Experimental Science and Applications in the Middle East." It is a bold and imaginative venture involving Bahrain, Egypt, Islamic Republic of Iran, Israel, Jordan, Pakistan, Palestinian Authority, Turkey and the United Arab Emirates. It may seem to some to be quite improbable that these states, many of whom have an ongoing history of disagreement and conflict, could agree on any common enterprise — but they have. Just as building a common accelerator facility at CERN united the recently warring factions in Europe in 1953, the proponents of SESAME have a vision that the facility may play the same role for the Middle East. Indeed it was at the suggestion of Herwig Schopper (see sidebar on Schopper), a former CERN Director-General, that SESAME, like CERN some 50 years before it, was created under the umbrella of UNESCO (United Nations Educational, Scientific and Cultural Organization).

The idea started in 1997 when SLAC's Herman Winick and DESY's Gustav-Adolf Voss suggested that BESSY I, the German synchrotron scheduled for dismantling in 1999, be reassembled in the Middle East. Germany gave the green light soon afterwards, and the SESAME project was born. In the summer of 2000, the site of the Al-Balqa Applied University, 30 km from Amman, was chosen to host the new facility. UNESCO gave its final approval to the project at the end of May 2002, and SESAME was established as an independent international organization in January 2003.

Meanwhile the components of BESSY have been dismantled, labeled, crated and shipped to Jordan. The enthusiasm for a synchrotron light source in this part of the world is intense. Although there are close to 50 synchrotron radiation sources in the world, very few are located in developing countries. This will be the Middle East's first synchrotron. Users in the participating states want to make sure it is of a high enough energy to produce the short wavelength and high brightness necessary to do cutting edge research. Hence building the new machine will not just be a matter of putting together a numbered kit of parts but will involve considerable redesign and improvement. This is quite a challenge in the context of Middle Eastern countries but is accompanied by vigorous technical training and technological transfer initiatives backed up with advice and encouragement from western nations. It is hoped that the machine should be up and running in the next few years.

Fig. 8.2 The King of Jordan discussing the SESAME Project, which will be located in Jordan and available to all scientists. On his right is ex CERN Director General Luciano Maiani and on his left ex CERN Director General Herwig Schopper, one of the prime movers in this venture.

Fig. 8.3 An aerial picture of the European Synchrotron Radiation Facility (ESRF) located in Grenoble, France. Construction was initiated in 1988 and "the doors were open for users" in 1994.

Fig. 8.4 Aerial view of SPring-8, a synchrotron light source located in Japan. Construction was initiated in 1991 and "first light" (an x-ray beam) was seen in 1997.

Sources. Examples are the Advanced Light Source in Berkeley or SPEAR III at Stanford. (See Fig. 8.5.)

VIII.7 Angstrom Wavelength Free Electron Laser Facilities

The fourth generation of synchrotron light sources may move away from the circular synchrotron towards a free electron laser (FEL) (see sidebar for FEL), invented by John Madey (see sidebar for Madey). In an FEL, a beam of tightly bunched electrons, accelerated in a linac, passes through a wiggler (similar to an insertion device for the synchrotons described above). The radiation emitted in the bends remains in phase with the bunches and, reacting back on the bunches, causes them to bunch even more sharply, enhancing the brilliance of the radiation. The device is called a Free Electron Laser because the effect is very similar to how the light from a laser stimulates the emission of further light in the laser medium: light that is coherent, adding up in amplitude so that double the amplitude produces four times the power.

The use of an FEL for a Fourth Generation Light source was pioneered by Claudio Pellegrini (see sidebar for Pellegrini). It is effectively an amplifier for x-rays but one cannot feed x-ray radiation into the wiggler;

there simply are no sources powerful enough. Instead, the first few wiggles are enough to produce some x-ray "noise," enough for the rest of the wiggler to amplify. This method, called "Self Amplified Spontaneous Emission" (SASE), was first produced in the microwave region at Livermore in the 80s; and then, at ever shorter wavelengths, at UCLA and at ANL. The SLAC Linac Coherent Light Source (LCLS), the European Union X-ray Free Electron Laser (EU XFEL) at DESY in Germany, and the SPring-8 Compact SASE Source (SCSS) in Japan are the new FEL light sources under construction at this time. These new facilities are based on very high-energy electrons (many GeV) and very long undulators (tens of meters). Figure 8.6 shows the Linac Coherent Light Source (LCLS) at SLAC. It employs 14 GeV electrons from the original SLAC linear accelerator. Making electron pulses of high peak current (more than a thousand amperes), with small size (transversely less than a tenth of a millimeter), and with little jitter, is a challenge, but appears possible.

Prior to constructing the European EU–XFEL, the DESY people have already constructed a VUV FEL — now named FLASH — operating in the vacuum ultraviolet. Until 2008 it will be the only FEL operating in the soft-x-ray range. This device is really just a 260 meter version of the XFEL (which will be 2.1 km

Fig. 8.5 The SPEAR 3 synchrotron dedicated source, located in SLAC at Stanford and first brought into operation in 2004.

Free Electron Laser (FEL)

The optical laser, which has become part of our daily lives, works by a process called "stimulated emission." Suppose a light wave, passing through a large number of atoms, or molecules, has the correct wavelength, frequency, and hence photon energy to match the energy interval between two quantum states of the electron system around the atom. If the upper level has been pre-excited, the photon energy (frequency) emitted when the atom falls back into the ground state will augment the intensity of the light wave. The wave will travel through the medium rather as a "Mexican wave" propagates around the terraces of a football stadium — coherently — that is, each contributor will be in phase with the arrival of the wave. The laser pulse that results from this coherence will be intense, monochromatic and consist of parallel wave-fronts with virtually no divergence — a perfectly parallel beam of light. Most materials used for lasers emit a fixed wavelength in the visible or near visible spectrum: a tunable source in the higher frequency domain of x-rays is badly needed, for example, for high resolution diffraction experiments on biological and crystal specimens.

The FEL is just such a tunable source and can provide high intensity radiation into the x-ray spectrum. The emission mechanism is rather different from a conventional laser but the coherence and amplification properties are similar. Suppose an electron beam and a light (x-ray) beam travel together through a series of regularly spaced bending magnets. We can imagine that synchrotron light is emitted at each bend and if it is both at the same frequency and in phase with the arrival of the light beam it will augment the intensity of the beam coherently. This will be the case for a particular wavelength depending on the magnet spacing. The array of magnets is rather like a diffraction grating.

There is another condition to be met. If the light beam and the electrons are not traveling at exactly the same velocity, they will slip in phase, one against the other. We must make sure that the distance needed for this slippage to reproduce the same phase relation between electrons and light in each magnet is equal to the magnet spacing. We can tune this condition and alter the wavelength of light produced by the laser by changing the bending strength and beam energy.

Photon emission can be triggered by an external laser beam traveling along the axis of the electron beam through the wiggler, or — particularly for energies above those available with normal lasers — by the "self-amplified" mechanism or SASE (Self-Amplified Stimulated Emission), in which the lasing "seed" is provided by a photon emitted by an electron in the early parts of the wiggler.

The reader may by now be wondering how the light beam can follow the electron beam as it is bent by all these magnets. The solution is to alternate the polarity of the magnets to form a "wiggler" so the electron beam does not deviate significantly from the straight path of the light.

Free electron lasers are capable of producing synchrotron light of an intensity and coherence and in a frequency spectrum beyond the present generation of circular synchrotron light sources. Major projects are underway in the US and Europe to build such an x-ray source and no doubt they will abound among the next generation of synchrotron light sources.

John Madey (1943–)

American physicist

When John Madey received a BS in physics from Caltech in 1964, his advisor was Alvin Tollestrup (see sidebar in Chapter V). He received an MS in electrical engineering from Caltech in 1965 under the pioneering laser physicist Amnon Yariv and went to Stanford University. There he worked, for his 1971 PhD, under Bill Fairbank to develop an intense source of thermal positrons for Fairbank's positron free-fall experiment.

While an undergraduate, Madey worked in the synchrotron laboratory on high-energy physics experimental equipment as well as on the synchrotron itself. He spent his summers at Brookhaven National Laboratory commissioning the AGS. In 1971 he joined the WW Hansen Laboratory at Stanford and during this time he invented, and then led a team that built, the first Free Electron Laser. He has said that his invention of the FEL was strongly influenced by a conversation he had, as an undergraduate, at Brookhaven with John Blewett, Ken Green and Renate Chasman in which they speculated about making electrons radiate coherently — just as they do in an FEL. In 1986 he was promoted to Research Professor in the Stanford Department of Electrical Engineering.

In 1989 he moved to Duke University where he led a successful effort to develop an FEL based on a storage-ring. In 1998 he became Professor of Physics at the University of Hawaii, applying FEL technology to remote sensing.

Madey has thought deeply about accelerator physics and he is on record for one particularly eloquent defense of accelerator physics that might well serve as a motto for this book: "accelerator (and FEL) physics are not isolated disciplines dedicated only to the construction of ever larger and more expensive facilities, but are an intrinsic part of the academic enterprise dedicated to the understanding of the most fundamental principles and phenomenon of electrodynamics. But since we are many times seen by our peers as the designers and enablers of the increasingly complex facilities they need for their research, we are, I fear, at risk of being thought of by our peers more as engineers and technicians than scientists. It therefore appears that, from time to time, we need to gently remind our colleagues that while we are happy to collaborate with them in the development of the facilities they need for their research, we have our own equally fundamental scientific and academic aspirations."

long). The VUV FEL uses electrons of 1 GeV and has a wiggler of 30 m length. (See Fig. 8.7.) In the future, they will construct the XFEL. This device will have electrons from 10 to 20 GeV, an undulator of 150 m length, and produce x-rays from 0.085 to 6 nanometers in pulses as brief as 100 femtoseconds. Because it will employ a new superconducting linac, the cost is much higher (in the $1 billion range) than that of the Linac Coherent Light Source (LCLS) (which uses the last 1/3 of the SLAC linac), but it will produce 30,000 pulses a second (unlike LCLS or the SPring-8 Compact SASE Source, Hyogo, Japan (SCSS), which will only produce 120 and 60 pulses per second). However, it should be noted that detector development is required in order to take advantage of this high pulse rate. Construction is expected to start in 2007, with commissioning in 2013.

The LCLS will provide x-ray pulses of 10^{12} photons of wavelength 0.15 to 1.5 nm, with a pulse duration of 100 femtoseconds down to 100 attoseconds and a pulse rate of 120 Hz. The average brightness of the x-ray beam will beat the present third generation light sources by a factor between 1,000 and 10,000. An ordinary dental x-ray installation would only have a "brightness" about ten million times weaker! The LCLS, due for completion by the end of 2008, will cost $300 million.

The challenges include the need to compress the electron bunch from 0.8 mm down to 0.023 mm, overcoming both space charge effects and coherent radiation effects, and the construction of a 122 m wiggler magnet to the necessary precision of field. The new FEL x-ray machines will open up a new world of science in biology, chemistry, and physics.

The SCSS will employ a new superconducting linac at 6 GeV and still reach x-rays of 0.1 nm because the wiggler wavelength will be only 15 mm. The wiggler gap will be very small (3 to 4 mm), but the length more than 100 m! Research was only started in 2004 (unlike the LCLS, which had been studied for a decade before construction was initiated in 2006). So, although the Japanese became interested after the other two groups, they are moving quickly and hope to have the project completed by 2010.

The Jefferson Laboratory is a leader in the development of FELs with a very high average power. High peak power is not unusual for an FEL, but high average power required the superconducting technology that Jefferson Laboratory had pioneered for CEBAF. In 1998 they exceeded previous records for FELs by a factor of 30 and soon, after considerable effort, they had reached 10 kW. For this they used a recirculating configuration containing, for the first time, an energy recovering linac. Efforts are underway to reach an even higher energy. It has been decided to move this facility to Los Alamos, where the initial goal is 100 kW. The 100 kW power level is below the "classified" category, but the US Navy supports the program, so this suggests that a

more powerful machine is being considered as a shipboard defensive weapon.

A number of other laboratories, including Cornell, are now studying the concept of a light source based on an energy recovering linac.

Claudio Pellegrini (1935–)

Italian-American physicist
International Free-Electron Laser Prize, 2000
American Physical Society R.R. Wilson Prize, 2001

Born in Rome he received the Laurea (1958) and then the Libera Docenza at the University of Rome in 1965. He was at the Frascati National Laboratory from 1958 to 1978, then at the Brookhaven National Laboratory, National Synchrotron Light Source where, from 1978 to 1989, he was Associate Chair and Co-director of the Center for Accelerator Physics. In 1989 he became a Professor of Physics at the University of California in Los Angeles (UCLA).

He is a theorist but, like most accelerator theorists, very well versed in the techniques needed to build such machines. His research has been centered in two areas. The first of these is the study of instabilities and collective effects in high intensity particle beams. The second is the study of the interactions of particle beams with electromagnetic radiation and plasmas.

He is well known for his work on instabilities including one that he discovered while working together with Matt Sands (see sidebar in Chapter VI) and Bruno Touschek during his Frascati days. This is the exotically named "Head-Tail Instability." Not only was he able to explain the phenomenon but he also developed a cure for it.

But it is in the second area of particle interactions with electromagnetic beams that he is perhaps best known, and particularly for devices that he invented and then deeply analyzed. In this second field, his studies have also paid particular attention to instabilities, collective effects, and phenomena of self-organization. One class of such devices employs lasers and plasmas to achieve very high energy. Pellegrini has always been in the forefront of efforts along this direction and since moving to UCLA he has been involved in experiments on the plasma beatwave accelerator and the Inverse Free-Electron Laser.

He has been particularly interested, actually since 1984, in the high gain, self-amplified spontaneous emission (SASE) regime of an FEL. He was the first to propose, in 1992, to use a SASE-FEL at the Stanford Linear Accelerator Center (SLAC) to produce high power, sub-picosecond pulses of 0.1 nanometer x-rays. This proposal has led to the Linac Coherent Light Source (LCLS) a 0.15 nanometer SASE-FEL. This major project presently under construction at SLAC may well prove to be the key to SLAC's future.

Fig. 8.6 Aerial view of the Stanford Linear Accelerator Center showing the Linac Coherent Light Source (LCLS).

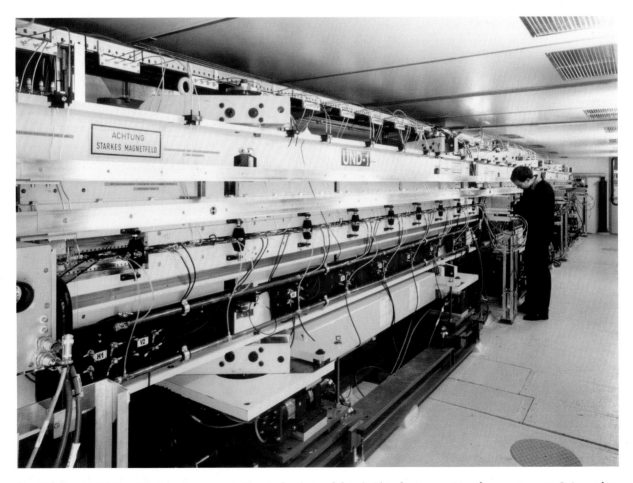

Fig. 8.7 The DESY Free Electron Laser magnetic wiggler (or undulator). This device consists of many magnets. It is used to produce laser light with short wavelengths; i.e., in the ultraviolet and x-ray regions of the spectrum.

VIII.8 Future Fourth Generation Synchrotron Sources

Over the last decade, or so, there has been considerable discussion within the community of users and builders, as to what would constitute further advancement in synchrotron sources; i.e., become the Fourth Generation.

Users seem to expect a variety of improvements. A higher flux of x-ray photons is one requirement, as is an even smaller spot size for the same number of photons, in order to enhance intensity. Other users would like ever shorter wavelengths (ever more energetic photons) while some want just the opposite: intense sources of 0.03 cm wavelength. There is also a group of users who would like to have even shorter pulses of x rays — well into the attosecond range.

From this list, it is clear that there can be no single type of fourth generation light source but a variety of machines. Certainly the third generation machines can be improved so as to produce more intense sources of x-rays. Such programs are underway, at Brookhaven, Argonne, Berkeley and SLAC; and new higher energy injectors can increase the number of electrons that can be stored. Other possible improvements include superconducting bending magnets (to decrease the wavelength of the x-rays), and continuous injection (to increase the available time of high intensity x-rays). In fact, at Brookhaven a major new synchrotron light source, costing about 800 M$ has just been authorized.

Another possibility is making special accelerators, or special x-ray beam lines on existing accelerators to provide THz radiation from bunches of electron of length 0.03 cm. An alternative for the generation of THz radiation is to make new, compact (table top) devices, employing laser acceleration.

One direction will produce copious amounts of very hard x-rays within a very narrow spectrum for the purpose of studying crystals and even biological samples *in vivo*. Two major facilities that will meet this requirement are the previously-discussed TESLA, or EU XFEL, in Germany and LCLS at Stanford.

Yet another possibility for the next generation, beyond these devices is a "green field FEL," being considered by a consortium in the US. It is hoped to reduce the wavelength of the x-rays by a factor of ten over the LCLS and for this purpose the electron transverse beam dimensions must be reduced by a factor of ten. In order to do this one might invest in R&D to improve the electron source, or to use a sophisticated manipulation of the electrons going into the wiggler. Another expensive route is to increase the energy of the electrons beyond

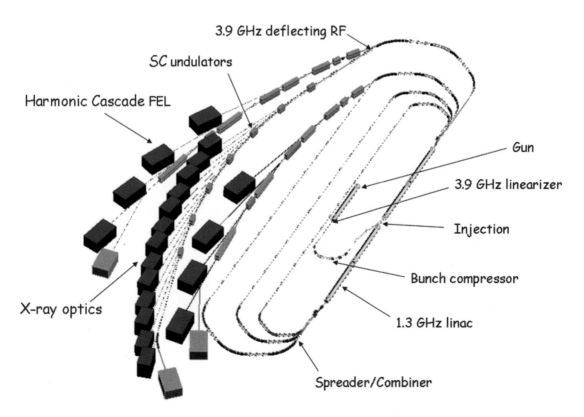

Fig. 8.8 A schematic of a possible fourth generation light source. This is the proposed facility LUX, as envisioned by a team at the Lawrence Berkeley National Laboratory.

the 15 GeV of the LCLS. Of course there will be more facilities built in the range of the EU XFEL, or LCLS. For example, the Koreans are making an FEL using a 3.7 GeV electron beam and an 80 m wiggler combined with many beam lines, constituting a fourth generation facility. It is also clear that pressure will develop for extending the range of FELs and increasing the number of beam lines for experimenters. Such pressure could, of course, take the form of improving the present facilities. In either case, there are challenges for the accelerator builders.

Another interesting possibility for a Fourth Generation Synchrotron Source is a more modest facility, but one with very short, femtosecond, pulses of intense x-rays — shorter than the timescale for the dynamic processes involved in the study of the dynamics of chemical reactions or melting of materials. The large facilities can also do that. However a smaller facility achieving the same short pulses, but at a longer x-ray wavelength, has interested potential users and more favorable economics. Such a device must be based upon a linac, rather than a synchrotron, so as to obtain very intense short pulses. The most economic form of linac is one of the re-circulating type in which beam energy is recovered.

Plans for such facilities, and the required R&D, are under way at a number of laboratories. An example, having particularly short pulses, and incorporating also an FEL, is the LUX Facility under study at Berkeley. It is shown in Fig. 8.8.

Chapter IX. Cancer Therapy Accelerators

This chapter is devoted to an important application of accelerators — medical diagnosis and therapy. A recent global survey of accelerators found that only 10 or 20 were high-energy devices feeding academic particle research, but there were more than 5000 smaller accelerators routinely used in the hospitals of the world.

There are two distinct ways — both important — in which accelerators can contribute to medicine. The first, upon which we will only briefly comment, is to produce isotopes for various medical diagnostic purposes. The second is to produce external radiation for the treatment of cancer. The radiation can be in the form of x-rays, produced from high-energy electrons, or it can be a beam of heavier particles: protons, neutrons, or ions such as carbon.

Each section of this chapter will deal with the beams from a particular type of machine. We shall not have a special section on electrostatic machines, although perhaps the very first use of a nuclear machine in a hospital was in 1937, when van de Graaff and John Trump constructed an electrostatic machine operating in the 0.5 to 1.2 MeV range. This was installed in the Harvard Medical School, where it produced x-rays for cancer therapy. Electrostatic devices have been superseded by betatrons and then by linacs.

IX.1 Cyclotrons

From the earliest days, Lawrence was interested in using his cyclotron for medical purposes. His first concern, however, was the harmful effect of radiation upon the people in his own laboratory. In the very early days they had shown a remarkable disdain for shielding and radiation protection. In the summer of 1935 Lawrence recruited his brother, John Lawrence, a medical doctor then at Yale, to review procedures in the laboratory. John Lawrence subjected a rat to the cyclotron beam, and the animal's death sobered the crew, until it was

discovered that the rat had really died from asphyxiation. Nevertheless, the visit resulted in vast improvements in safety procedures.

Lawrence and his team were quick to realize the application of accelerators to produce radioisotopes. His cyclotrons produced isotopes as early as 1934, and by 1937 the cyclotron was running 24 hours a day to make isotopes, which Lawrence provided free to hospitals around the world.

Nowadays radioisotopes are mainly used in many of the medical imaging techniques which are available for diagnosis. A radioisotope of technetium is particularly popular. It emits a characteristic soft gamma ray of 141 keV and it has a convenient lifetime of six hours. Technetium was in fact the first element to be artificially produced (1937). This isotope is also commercially produced in nuclear reactors (often of the kind that uses highly enriched uranium suitable for making a bomb).

Some of today's advanced hospitals have a cyclotron in the basement for producing isotopes which are then delivered throughout the hospital by compressed air tubes. Speed of delivery is important, for the isotopes used for diagnostic purposes must have short lives so as not to remain in the body long enough to cause permanent damage to the patient.

Radiation from longer-lived radioisotopes, such as cobalt-60, is used in the Gamma Ray Knife, developed in the 1980s at the University of California, Los Angeles (UCLA) and used as an alternative surgical tool, especially for brain surgery. A later linac-driven device is now commercial and is made by Radionics and Electra, Inc. The first such Novalis unit was installed at UCLA in 1997.

Still another use of radioisotopes for treatment rather than diagnosis is to implant radioactive "seeds" in the cancer. This "brachytherapy," for example, is used as one method of treating prostate cancer and competes favorably with surgery and external radiation treatment.

We turn now to the second and most extensive use of accelerators in medicine — the production of external radiation employed in cancer therapy. We recall that even as early as 1933, Lawrence supposed that fast neutrons (generated, for example, by break-up of deuterium) would be better at destroying cancers than x-rays. At that time, Lawrence's graduate student David Sloan made a 1 million volt x-ray tube that even as early as 1933 was being used by radiologists at the University of California medical school in San Francisco to treat cancer patients.

In 1937, Lawrence's mother was diagnosed as having an inoperable uterine tumor. Treatment with Sloan's x-ray tube seemed to cause the tumor to disappear, and although there is question as to whether or not the diagnosis was correct, the Lawrence brothers were highly motivated to push ahead with neutrons, which they believed would be even more effective. In the summer of 1936, John Lawrence drove from Yale with dozens of cancerous mice for experimental treatment with the 37-inch cyclotron. Soon after, a clinical program was started under the direction of John Lawrence and R.S. Stone — both medical men — and some, but as it turned out not enough, preliminary trials were performed during 1938 and 1939. The complete clinical program involved 226 patients but they were given too much radiation and their long-term prospects for survival turned out to be very bad. When the program was reviewed in 1948 only one of the 24 patients treated at the 37-inch cyclotron was still alive, and only 17 of the 226 treated at the 60-inch cyclotron were still alive. With hindsight it was judged that this was disappointing and x-ray therapy would have been far superior, and these poor results put an end to neutron therapy for more than 20 years.

The use of neutrons in therapy was revived around 1970 at the Hammersmith Hospital in London. They used a cyclotron and in the first decade treated more than 3,000 patients. Here the treatment was restricted to superficial tumors. A neutron radiation facility is presently in operation at Fermilab using a proton beam deflected out of their linac injector and producing neutrons from a beryllium target. Neutron therapy is also practised at the Harper-Grace hospital in Detroit.

The advantage of using charged hadrons (protons and other heavy particles) rather than neutral particles was first appreciated by Robert Wilson in 1946. Subsequently therapy with protons was pioneered at both Harvard and Berkeley. Although the first clinical examples of proton treatment were performed by John Lawrence in 1954 using the 184-inch cyclotron, the longest continuous history of the use of protons has been at Harvard, where more than 7000 patients have been treated since 1961.

The essential advantage of protons over x-rays and neutrons is that the radiative energy is deposited at a particular depth within the person, in a thin layer called the Bragg peak. The depth at which this occurs depends on the energy of the hadron. In contrast, x-rays deposit much of their energy in the tissue of a patient close to the surface and then less and less as they penetrate towards a deep-seated tumor. It is therefore difficult to kill the cancer but not the healthy cells on the way to the tumor and beyond it. To spare healthy tissue, it helps to direct radiation onto the tumor from many different directions. Although this can be done with x-rays, it is most effective with hadron beams when it may be combined with the control of energy deposition depth that the Bragg peak allows. The Bragg peak is particularly effective at sparing a delicate organ just beyond the site of the tumor.

The method of directing the beam on the tumor from many sides employs a gantry, or moveable arm of magnets, between the accelerator and the patient. The gantry rotates about the patient and delivers the beam to the same precise spot from many angles while the patient remains stationary. The complex, expensive and heavy components of the gantry must be hidden from the patient, who might become frightened at being at the center of a massive piece of moving machinery.

A number of small cyclotrons (60–70 MeV) have been used with considerable success in treating eye tumors. In recent times IBA (Belgium) and ACCEL (Germany) have produced cyclotrons in the 220 MeV range dedicated to proton therapy in hospitals. The first machine from IBA was installed in Massachusetts General Hospital while ACCEL has built two superconducting cyclotrons — one for PSI in Switzerland (not a hospital) and the other in Regensburg outside Munich. About 40,000 patients have been treated with proton beams, and the companies involved in the commercial production of proton machines include ACCEL, IBA, Hitachi, Mitsubishi, and Optivus. Recently Siemens, who participated in the construction of the Heidelberg machine, have also joined the party.

IX.2 Linacs

In spite of the advantages of hadrons, most patients with cancer are treated, and treated very well, with x-rays. The method of producing the x-rays is the same as that discovered by Roentgen in the late 19th century and used in the original Crookes tube and its descendants in every dentist's surgery. Electrons are accelerated in a linac and then shot into a metallic target to produce x-rays. The microwave linacs that are to be found in many hospitals range in energy from 4 to 22 MeV. They are simple, compact, require virtually no technical maintenance and are rather inexpensive compared with the cyclotrons or synchrotrons needed for hadron therapy. Varian and Siemens are the principal companies involved in their manufacture, but some of the companies mentioned above also make linac-x-ray machines. The linac, followed by a vertically deflecting magnet system that can turn the beam through 90 degrees, is small enough to rotate about the patient, forming a rather simple form of gantry, as can be seen in Fig. 9.1. The x-ray-producing linacs are highly developed machines. These machines are reliable, compact, safe, easy to use, and highly programmable so that the x-rays can be tailored in both intensity and spatial extent for each patient.

Since most large hospitals can afford these x-ray-producing linacs, the patient can remain in a local hospital and be near family and relatives while under the care of their normal physician and a local oncologist.

Treatment by hadrons would sometimes be better for some cancers, but hadron therapy apparatus is too expensive for all but the largest of national treatment centers, and then only in developed countries including Japan, Taiwan, Korea, China and South Africa. The obstacle seems to be the initial capital outlay, for it can be argued that in the long term an investment in an accelerator for therapy is cheaper, per patient, than other widespread treatments like chemotherapy — even when you include the cost of servicing the loan to build it.

IX.3 Synchrotrons

Synchrotrons (and cyclotrons) of about 250 MeV are used for proton therapy. This energy is sufficient to reach any internal organ. Synchrotons are presently the preferred method of accelerating ion beams of similar penetrating power. Conventional cyclotrons with massive poles would be too big, although fixed field alternating gradient (FFAG) cyclotrons, with their compact ring of components, are now being considered for this purpose. The synchrotrons used at Loma Linda in

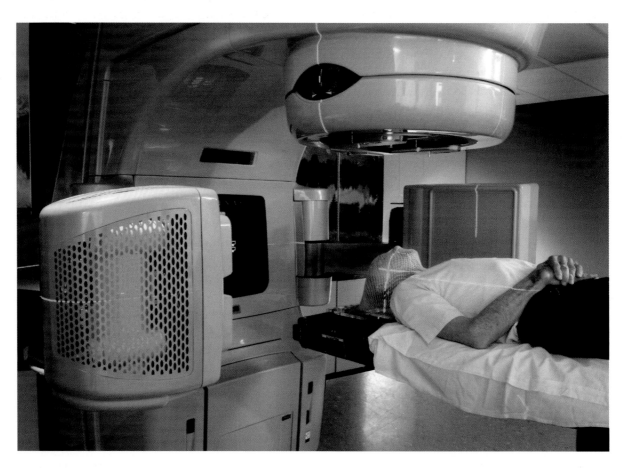

Fig. 9.1 A modern system for treating a patient with x-rays produced by a high energy electron beam. The system was built by Varian Medical Systems and shows the very precise controls for positioning of a patient. The whole device is mounted on a gantry. As the gantry is rotated, the electron beam, and the resulting x-rays, can be delivered to the tumor from all directions.

Los Angeles; and the Heavy Ion Medical Accelerator Facility (HIMAC) in Japan, were the first to follow the early hadron work at Harvard and Berkeley. The alternative of using a superconducting cyclotron for protons has led to recent projects in Massachusetts, USA, and National Accelerator Centre in South Africa. Cyclotrons, with their continuous beam of small emittance, enable very precise control of dose and treatment zone but energy is not variable.

Altogether there are now proton facilities at five hospitals in the US. Following Loma Linda, the second is the Northeast Proton Therapy Center, which started operation in 1999 and is located at Massachusetts General Hospital. The third is the Midwest Proton Radiotherapy Institute at Bloomington, Indiana, which is just coming into operation as we write, while facilities are being developed in Texas and in Florida. More proton facilities are being planned in the US, and many others exist or are being planned around the world.

Studies on the use of accelerated light ions (carbon and oxygen) were carried out in the 1970s and 1980s at Lawrence's laboratory in Berkeley, using a combination of a linear accelerator (the HILAC) and a synchrotron (the Bevatron) to form the Bevalac. In many ways ions are preferable to protons. Like protons, ions deposit their energy in the Bragg peak at the site of the tumor but the density of the deposited energy increases as the square of the atomic number, making heavier ions even more effective than protons. Research into the density of ionization as ions pass through cells suggested they are better suited than protons to "taking out" large sections of DNA, which cannot be repaired. Basic biology studies have led to an understanding of the dose delivered and the cancers most readily cured by ion treatment. They showed that carbon was the most suitable ion. The dose concentration of carbon is roughly 36 times better than that from protons. However, heavier ions, which might increase dose concentration even further, tend to fragment as they enter the patient, and penetrate beyond the target. The Bevalac facility was used to treat many patients (although not, of course, in a hospital setting).

This work prompted the Japanese to build the very large accelerator facility based on the use of light ions accelerated in a synchrotron in the Chiba prefecture (HIMAC, see Fig. 9.2). Work started on constructing this facility in 1987, it came into operation in 1994, and in its first twelve years of operation it treated more than 3000 patients.

Another ion installation is at the GSI Laboratory in Germany. Like the Bevalac, this was an experimental facility that found its place in a laboratory mainly devoted to nuclear physics research. Europe meanwhile is constructing or planning a number of new hospital-based facilities that concentrate on the use of ions. The

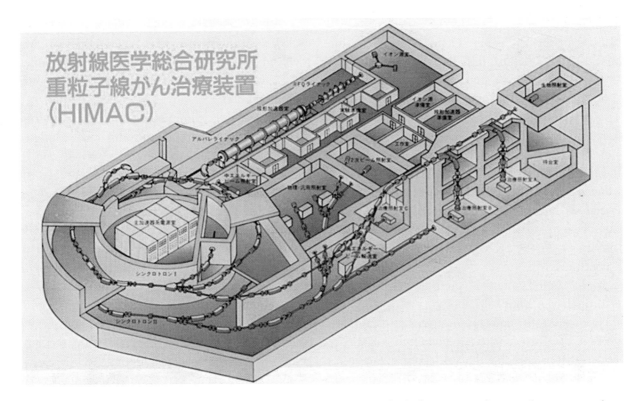

Fig. 9.2 A drawing showing the Japanese proton ion synchrotron, HIMAC. The facility consists of two synchrotrons, one above the other, to maintain a continuous supply of ions (or protons) to the treatment area. The pulse of ions is synchronized with the respiration of the patient to minimize the effect of organ movement.

Heidelberg Ion Therapy Centre (HICAT) — conceived in a collaboration between GSI and the Heidelberg Centre for Tumor Diseases — was approved in 2001 and civil engineering work started in 2003 aiming at treating its first patient in 2007. The second European centre, an initiative of Ugo Amaldi, is the *Centro Nazionale di Adroterapia Oncologica*, now being built in Pave, Italy. Other carbon ion centers are under consideration, such as the Med-Austron in Wiener Neustadt, ETOILE in Lyon, ASCLEPIOS in Caen, and a centre proposed by the Karolinska Institute in Sweden. Other synchrotron projects that include the ion option are planned in Japan and in Italy. Following up on the Berkeley and Chiba success, European work is based upon the pioneering work of the "Proton and Ion Medical Machine Study" (PIMMS) carried out at CERN by a working group with members from various institutes.

Although more facilities are under construction in many countries there are strangely no plans for a carbon facility in the US. This may be because the US government agencies rely on private industry to construct these machines and industry have focussed on protons. Since the Bevalac no longer exists, US victims of cancers best treated with light ions might ask to go abroad for their treatment.

IX.4 Other Therapies

Pions have been used as therapeutic particles at the LAMPF linac in the US, at the TRIUMF cyclotron in Canada and at PSI. A pion deposits very much more energy when it disintegrates (into an electron and neutrinos), than does a proton or a heavy ion but over a much less precise region. Early clinical trials, prior to 1994, treated advanced prostate cancer and glioblastoma, but showed no additional benefit of the therapy. Careful clinical trials, involving 300 patients, have been initiated at TRIUMF, where the results still await careful analysis, but so far show no clear therapeutic advantage of pions (which are very much more expensive to produce). At the present time pion therapy has been discontinued.

Finally, we should note that there has also been a study at Brookhaven National Laboratory of a different application of neutrons; namely boron neutron capture therapy (BNCT). This consists of giving to a patient boron attached to a molecule that seeks out the cancer. The patient is then bombarded with slow neutrons, which the boron absorbs and then releases radiation which is localized at the site of the cancer. Although this is theoretically appealing, the results so far have

not been encouraging. To date, most studies have used reactor neutrons, but accelerators offer many advantages in cost, size, safety, and the ability to produce slow-moving neutrons. If trials expand and prove effective, accelerators would no doubt become popular as the neutron sources.

IX.5 Future Facilities

As we have seen, accelerators have been used for cancer therapy since the days of the earliest cyclotrons. The current need is for a machine, together with its gantries and infrastructure, that a large number of hospitals can afford and yet which can operate with minimum technical backup. The beam intensity needed is small but there must be precise control of the beam intensity, energy, and position. The dose of radiation delivered must be very precise: spread evenly over the tumor and must spare healthy tissue, which may be as close as a millimeter or two from the target zone. The medical community seems to accept that a proton machine of 200 MeV/amu, or an accelerator of light ions of twice this energy, is the ideal prescription. Lower energy protons are useful for tumors which are close to the surface (e.g., in the eye).

Accelerators producing x-rays (linacs) will continue to be built at the rate of a few hundred a year. As far as hadron therapy (protons and ions) is concerned, we can expect that there will be an increasing number of cyclotrons and synchrotrons. Among the five companies involved in producing proton therapy machines, IBA and Siemens are studying how they might enter the ion therapy market.

The research community is also considering new accelerators for therapy and different configurations of old methods. A Japanese group as well as a Russian team from Novosibirsk, for example, are considering the "cooling" of the ion beam to reduce the size (and cost) of the beam delivery gantries. Another approach to the same problem is to improve the ion source so to make beams which are much smaller initially. A third possibility is the use of a new kind of cyclotron (called a non-scaling FFAG) which has special advantages. A fourth possibility is the use of superconducting magnets so as to greatly reduce the size and weight of the gantry beam delivery system. (The Heidelberg facility has a gantry system that weighs 600 tons.)

A German-American team is studying yet another possibility: the use of small superconducting cyclotrons which can rotate around the patient, avoiding the use of a complicated and costly gantry. They plan to use a spiral ridge cyclotron, a form of FFAG, well-understood and even produced commercially.

Accelerator therapy is certainly widely used in the developed countries, for it is medically effective, and compares favorably with other forms of treatment in cost effectiveness. Thus, improving these facilities and, even more importantly, bringing these cancer therapy machines within the grasp of more people of the world will clearly be a major activity of accelerator builders in the years ahead.

Chapter X. Past, Present and Future

In this book we have traced accelerator history over nearly a century that has seen the construction of bigger and better machines for particle physics research. Now the time has come to look ahead. Activity in the field of particle accelerators promises to be even more vigorous than the past; but as Niels Bohr said:

"Prediction is very difficult, especially if it's about the future."

The primary aim of the scientists and engineers who built these machines was to accelerate to higher energy in order to probe deeper into the structure of matter. At the same time they found a multitude of practical uses for these machines. In this chapter we will look at the future of accelerators for these applications as well as for the "high energy frontier" of particle physics.

In Chapter VIII we described electron rings of more modest energy built as sources of synchrotron radiation. Though the energy was small, great effort was put into making intense beams of tiny diameter, this brilliant light gave the sharpest diffraction patterns and other benefits for the users. High intensity at a modest energy has also been the aim of another kind of medium energy accelerator — the spallation neutron source.

Accelerators are the enabling technology for many applications besides those two. There are some 200 high- and medium-energy machines and probably as many as 16,000 smaller accelerators. About 1,000 of these are used for low-energy studies; 200 are to produce radio-isotopes for medicine; almost 8,000 are employed in the therapy of cancer, and another 8,000 for industrial processing, ion implantation, modifications of surface and bulk materials and for sterilization of food, for example, or for killing anthrax spores in mail. These machines are of very low energy and typically produce intense beams of electrons of only 6 MeV to be comfortably below the Coulomb barrier value at which electrons can leave materials that they strike radioactive. Most are

manufactured commercially; one firm produces over 500 medical therapy accelerators each year. Although these accelerators are not at the frontiers of high energy and large size, many are at the technological frontier, demanding performance that can only be achieved with innovative ideas. For example, medical machines, besides needing to be reliable, require sophisticated systems to control beam energy, intensity and spot size in order to ensure the patient's safety.

X.1 Future Needs

A series of landmark technical breakthroughs in accelerator design and construction have ensured steady progress towards high-energy at the frontiers of energy, intensity, and beam quality over the last decades. These include alternating gradient focusing, the use of superconductivity for magnets and accelerating cavities, sophisticated electron and ion sources, an understanding of how to store and collide beams, elegant control systems, and the invention of cooling as a means to compress beams. Each new generation of accelerators, especially those employed in high-energy physics, have been more efficient and economic than their predecessors. In spite of such progress, each has been substantially more expensive, to the point that new projects now find themselves in competition for funds with — for example — a space initiative, a major activity in some other branch of physics, nano-technology, advanced computing, or the field of microbiology. Inevitably they are perceived by scientists in other fields as a threat to their own budgetary ambitions. Already we have seen one major project, the SSC, cancelled because of its effect on the balance of the science budget of the world's richest nation (see sidebar on the SSC in Chapter VI).

Before reviewing prospects for the future we should see what the various high-energy and nuclear physics laboratories need. Their dreams may prove too expensive, but unless there are dreams there will be no progress. Fermilab seeks an ever-increasing luminosity through electron cooling, improved stochastic cooling and better stacking. They also hope to build a new, intense, proton driver, perhaps an 8 GeV superconducting linac, as the front end to their Tevatron complex. This would allow them, amongst other things, to increase the intensity of neutrino beams. SLAC has plans to upgrade their asymmetric collider, and KEK also has similar plans. Brookhaven has hopes of adding an electron ring to RHIC and thus being able to study very high-energy electron-ion collisions. CERN is already thinking of improvements to the LHC, such as increased luminosity and ion-ion collisions. This ferment of activity is just what one might expect to see from laboratories on the forefront of intellectual advance.

The major problem to be faced is that electron and hadron ring colliders seem to have reached their ultimate size in LEP and LHC. The next high-energy project, an electron collider, cannot be a ring but must revert to acceleration in a straight line; for this linacs must be made far more effective in terms of GeV per meter. After a lot of soul searching, the high energy community has homed in on the International Linear Collider as the future hope of particle physicists. In the past the solution for nations that cannot afford bigger high energy machines has been to band together the budgets and aspirations of several countries. Now, in order to build the ILC, continents must merge their resources.

In the even more distant future there will no doubt be further steady improvement in the techniques of linac or synchrotron design, but it is unclear whether these will be sufficient. It is very probable that we shall have to radically extend the accelerator builder's repertoire by seeking a new kind of device. Later we will review the devices and methods that might be expected in the future, but for the moment let us return to what is planned in the next year or two.

X.2 Linear Colliders and Their Origins

Both LEP and SLC were electron-positron colliders: the result of a swing of the pendulum from protons and antiprotons to leptons. The pendulum's swing followed the success of the antiproton-proton collider that had first created the W and Z bosons. Once the euphoria of the discovery was over there was a need to collide electrons and positrons for more detailed studies. We explained in Chapter VI that leptons, being point-like particles, are ideal for pinning down the precise properties of the newly discovered bosons. Hadrons, such as protons and antiprotons, though easier than electrons to accelerate to high energy, are rather blunt instruments of science. They are bundles of three quarks of different kinds, and their collisions are pictured as encounters between two of the six quarks involved: an ambiguous situation when it comes to interpreting the results — and a much lower energy of effective collision than one might think. Consequently, the lower energy of lepton colliders is not as serious a problem as it at first seems, for only about ten percent of that carried by proton and antiprotons is available when the two quarks collide — not so for leptons. In fact two of LEP's 100 GeV leptons are as effective as a proton and antiproton colliding in the Tevatron at close to 1 TeV.

We also explained that inevitably, once LEP had completed its precision work, the pendulum had to swing back again as the brute force of a hadron collider of even higher energy was once more needed to search for a new and heavier particle: the Higgs — a massive particle and the key to explaining the strange variety of masses among fundamental particles. CERN is busy building this new proton collider: the LHC, which will have an energy of 7 TeV per beam. However, new machines take several decades to develop, and, even before the LHC begins operation, one must prepare for the day the pendulum swings back to the use of electrons to reveal the subtle detail of the Higgs.

Of course the pendulum is an illusion, and the decision lies in the hands of the world community of particle physicists and accelerator builders — not to mention the governments of the many states who must band together to pay for the next step. The community is diverse and many interests come into play. Research and development programs investigating new methods spawn workshops and international meetings to recruit the worldwide interest of accelerator scientists to their cause. The International Committee for Future Accelerators (ICFA) considers these various interests.

Naturally, there is a tendency for the continent that failed to host the last large machine to expect the next in order to sustain its own community of experts, and bring the research to be done by its PhD students closer to home. For example, it is strongly argued that the US should make a bid to site the next linear collider to maintain high energy physics in the USA. Both the Tevatron and the B-Factory are expected to be closed within this decade, and there is a real danger that the pursuit of high-energy particle physics in the US will then rapidly decline.

Laboratories that have acquired experience in one particular kind of accelerator or collider — whether electron or hadron, linear or circular — hope that their expertise is used. The problems of hosting a truly international laboratory with ease of access by nationals of all member states, and with a future protected from unilateral budgetary decisions among these states, must also be addressed.

As the time approaches for governments to be asked to stump up the money, ideas have to crystallize into feasibility studies and design reports stuffed with cost estimates and reports of test facilities constructed to verify the principles of the design. A Project Leader has to be chosen and, no doubt, many years will be spent determining where the machine is to be sited.

Accelerator builders are never quick to admit that they are unable to build the next higher energy machine simply by scaling up and improving the cost per meter of magnet, tunnels and other equipment. But LEP is really the end of the line for circular electron colliders: the next electron machine must be a very different animal. Just imagine scaling up LEP. Even though it was given a circumference of 27 km to minimize synchrotron radiation losses, the leptons lost 3 per cent of their energy on each turn. Such synchrotron radiation losses stem from the fact that one must bend in a circle. They rise, not just with the square of the energy, but with the "square of the square." Scaling up LEP to a "modest" 250 GeV would send the loss up by a factor of 40; leptons would not even complete one orbit. Scaling up the huge circumference, from LEP's 27 km, might help; but to restore the loss per turn to LEP's 3%, one would have to increase the circumference of the new machine to 1000 km. Funding agencies might think of better applications for a tunnel of this length — and of course we forgot to say that we need not 250 GeV per beam, but perhaps as much as 500 GeV per beam.

The accelerator community therefore must turn to the linear collider. Compared to SLC, it will need ten times the energy and much more intensity. It will also be based on two opposing linacs rather than the single linac and two collider arcs of the SLC. Even though SLC has in a general fashion shown the way, much work needs to be done. To develop such a radical concept needs many years of design studies, feasibility studies, theoretical studies, and proof of principle prototypes before one may dare to ask for money to build it. As in all such studies, the several hundred or so engineers working on the ideas have had to patiently work their way through a maze of technical options, many of them leading only to blind alleys, before they can be sure they have a proposal with a credible time-scale and a cost estimate that will not soar out of control. Fortunately much of this is already behind us, but it is well worth recalling this activity to show how accelerator designers prepare their projects.

The late 1980s saw a burst of activity in a number of laboratories towards a linear collider, but it soon became clear that building one would not be easy. There were two major problems to be overcome. The first was the sheer amount of electrical power to be provided. Each electron carries only a tiny amount of charge: 1.6×10^{-19} coulombs. Multiplied by the number of particles in the beam necessary to get a good luminosity, say 10^{31}, this is 1.6 microcoulombs. Further multiplied by the voltage of 250 GeV (2.5×10^{11}) we find we need to supply 3.2 megajoules of energy for each "shot." With a rapid-fire repetition rate (say 100 shots per second) necessary to produce a useful luminosity, this becomes 320 MW of power — a large fraction of a new power station! This is fundamental and an inevitable consequence of the fact that in a linear collider particles have to be accelerated from rest and only get one chance to collide — they do not come round again.

Naturally, people not only played with the numbers, but thought of different concepts. Because the beam need not be kept in good enough shape for recirculation, it may be focused down to a much smaller spot at the collision point than is possible in a circular machine. The smaller beam size means more intense beams, and a greater number of interactions per shot, and thus more user data. The powerful interactions between the two beams actually produce a sort of pinch, which helps a bit. At this stage linear collider designers entered a long period of grinding through many alternative combinations of intensity, repetition rate, the number of bunches in each shot, the beam size etc. to convince themselves, their lab directors, and finally the guardians of the taxpayer's money that the design is the very best that can be devised (with the present degree of understanding and state of technology).

One of the important steps in getting funding for a project is to ask the physics community — the users — if it is really what they want. In the formative years of the linear collider there have been a multitude of conferences — "workshops" as they have come to be called — when the machine designers, primarily physicists, lay out their plans and particle physicists look at what they need in order to "do physics" using the machine. Inevitably at such gatherings the users point out that, in order to see more clearly the fine detail that a higher energy collision can potentially reveal, they will need even greater luminosity; the beams must be more intense and, particularly in the case of a linear collider, more power will be needed to accelerate the beam; and the spot size at the collision point will need to be even smaller. The reason for the users' appetite for luminosity is, of course, the wave nature of particles: their effective size shrinks with energy and the probability of an interesting collision diminishes. This might be compared to looking down a microscope: turning the turret to increase magnification often requires adjusting the mirror for brighter light.

Rather early on, linear collider designers realized that not only was the power that has to be fed to the beam of the order of hundreds of MW, but at least as much power again is wasted in the klystrons. The walls of the linac cavities carry strong currents, and a lot of power is wasted in ohmic heating. At this point it seemed to the embattled proponents of the linear collider that they would have to buy more than one complete new power station — and pay its running costs! The only solution was to refine the design and to use clever engineering to reduce the losses and bring the cost per meter of these high-tech components within a credible budget.

It was realized that the most effective way of reducing the power needs of a linear collider is to increase the frequency and thus scale down the transverse size of components. A higher frequency cavity can sustain a higher accelerating gradient reducing the length of the linac, the energy stored per meter, which must be provided by the power source, is smaller and the ohmic dissipation in the walls is smaller; factors that all reduce the power bill.

However, this scaling down brings its own problems: as the lateral dimensions of the cavities shrink to save stored energy, the conducting walls of the cavities are much closer to the beam. The electrical images of the beam in the cavity walls can send the beam into wild instabilities and limit both beam current and luminosity. This particular problem took a long time to solve. Cavities with slots to bypass away the unwanted modes of field and detuning (in which you make sure that there are slight differences among the cells of the cavities so that they do not resonate with the beam) had to be understood. These were simulated with massive computations and tried out in prototype cavities fed with beam from the SLAC linac.

Accelerator physicists had also to face up to the sheer length of the linacs. We know from SLAC's two-mile linac that 50 GeV needs at least 3 km of linac cavities, whose average field must therefore be 15 to 20 MeV per meter. A 250 GeV linac built in the same way would be 15 km in length, or 30 km if you include the positron linac facing the other way. This is about the same length as the LEP tunnel and might be a feasible aim, but for the fact that higher energies, possibly as much as 500 GeV per beam, may be needed to study the Higgs. One way to increase the average accelerating field is to scale down the cavities from SLAC's 3 GHz to the much smaller (by a factor of four) cavities around 12 MHz. The difficulty with this is that the power supplies — klystrons — also have to be scaled down, and the strong electron beams that link the input and output cavities of these amplifiers begin to suffer from instabilities. All this took a long period of engineering research and development.

Meanwhile at CERN — whose immediate needs will be satisfied by LHC and who therefore can afford to take a longer perspective — chose to carry this logic a step further and raise the frequency to 30 GHz, where there is no hope of building a conventional klystron. Instead they aimed for a solution where the linac would be powered by a second, drive linac with a low energy but high current beam. This is placed alongside the main linac, and 30 GHz power is tapped from cavities in the drive linac through a transfer structure to power the main linac's cavities. This scheme was christened the Compact Linear Collider or CLIC.

In situations like this different teams in various centers around the world naturally tend to have their own original ideas about the best solution, and a certain amount of friendly rivalry develops. This has the effect of pushing things forward provided the rival schemes do not become an impediment to collaboration on common technical issues. Accelerator labs, in contrast to industrial firms and some research establishments, seem to be particularly good at this balancing act.

SLAC, calling its new machine the Next Linear Collider (NLC) and KEK in Japan (the JLC), concentrated (after rejecting a collider running at a lower frequency) on room temperature cavities at 12–14 MHz. At these higher frequencies, the problems of reaching the theoretical increase in accelerating gradient seemed not as easy as the mathematics might have suggested. Cavities began to break down as "dark currents" eroded their surfaces. Ways of coaxing, or conditioning, cavities to reach their theoretical performance by gradually easing up the voltage had to be developed.

The worldwide effort involved a number of experimental programs of major size. A final focus test area was developed at SLAC, and a method for measuring the very small transverse size of beams was developed by Tsumoru Shintake and then demonstrated at SLAC. A test accelerator at the higher frequency of 12 GHz was built at SLAC (see Fig. 10.1), and a test damping ring was built at KEK (see Fig. 10.2), while the CERN team worked on their approach by building the CLIC test facility (see Fig. 10.3).

Meanwhile DESY in Hamburg, having constructed the first superconducting collider in Europe (HERA), was eager to follow its own ideas: the project was called the TeV-Energy Superconducting Linear Accelerator (TESLA) (see Fig. 10.4). TESLA's director, Bjorn Wiik, in partnership with Maury Tigner of Cornell University in the USA, saw superconducting cavities as the way to reduce the huge power bill by eliminating ohmic losses in the cavity walls. Superconducting cavities, although more expensive per meter than their room temperature alternative, have the advantage that they do not have to be filled with energy every time a linac pulse comes along. They can be left "switched on," and the linac only has to deal with a relatively thin trickle of more uniform beam, instead of the very short, very intense bunches appropriate to the room temperature linac. The nub of the question for TESLA was: how to build a superconducting cavity cheaply, harnessing all the advantages of mass production to make this rather alien technology cost effective.

Apart from the new technology, a collider needs a positron source and two small but intricate synchrotron rings to prepare the beam before it is accelerated towards the high intensity collision point. These damping rings store the electron and positron bunches, giving them a chance to lose energy via synchrotron radiation in such a way that the side to side motion of the particles and their energy spread shrink, a very effective cooling mechanism. Here the SLC experience was very important.

X.3 The International Linear Collider (ILC)

In this rather lengthy account of the linear collider we have tried to give you a flavor of the scale of the endeavor. Members of many teams — some physicists, some engineers; some using pure theory, others building prototype and test stands — work for many years communicating their progress through publications and conferences until the project is ripe for a financial decision by a country, continent or international funding agency. As we write, the situation for the ILC is that the harvest is in sight. A panel of expert consultants has

Fig. 10.1 The X-Band Test Accelerator at SLAC. Here one of the approaches to an International Linear Collider was tested by actually building a section of a collider. The test was highly successful.

Fig. 10.2 The damping ring built at KEK, Japan, in order to study the process of making a beam of very tiny dimensions, as would be needed for the International Linear Collider. The damping ring is shown when it was under construction.

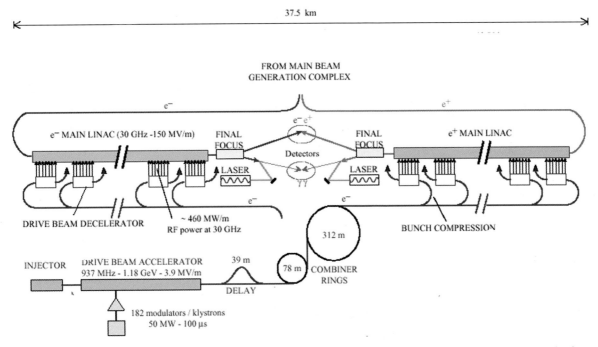

Fig. 10.3 A diagram showing the CERN approach to a linear collider. The two main linacs are driven by 30 GHz radio frequency power, derived from a drive beam of low energy but high intensity that will be prepared in a series of rings combined with a conventional linac.

Fig. 10.4 TESLA technology: these superconducting accelerator structures built of niobium are components of the future linear accelerator.

recommended that the first priority should be to prepare a project, an International Linear Collider, based on TESLA and its superconducting technology. A managerial structure embodying the collaboration of all the particle physics institutes that were previously working on TESLA, NLC and JLC has been set up. It has been decided to follow the superconducting path and to proceed with the fascinating exercise of choosing a place to put the new machine among the continents of the world.

A Director (Barry Barish) has been chosen, discussions on the choice of site are under way and the sequence of events; R&D followed by optimization of design and component details leading to a final proposal; is growing apace. All of the major laboratories are involved in the R&D, with their efforts coordinated by Barish. Already a gradient of 35 MeV/m has been chosen, which would, in a realistic length (30 km), lead to 500 GeV per beam at a pulse rate of two thousand times a second. The energy is a decision that may have to be revisited when the first results arrive from LHC. As we write, it is not clear whether we can expect to

find one Higgs, probably within the reach of such a machine, or a family of Higgs particles whose properties may suggest a whole host of supersymmetric particles at even higher energies.

Before moving away from the ILC we should mention a couple of possible "bells and whistles." Although the electron-positron collisions are of primary interest to particle physicists, electron-electron collisions open up different "channels" for interactions that allow studies that are not possible with electrons colliding with positrons. The added cost of this capability is trivial.

The second possible extension is even more interesting, for it opens up very different and important channels for particle physics: namely to have collisions between gamma rays! Lasers are very powerful but produce their intense light mainly in the visible spectrum; certainly not in the region of gamma rays. But suppose one points the intense beam of photons from a laser at a high-energy beam of electrons. Some — perhaps one in a billion — come bouncing back with higher energy than the electrons; i.e., as gamma rays. Lasers can readily supply the number of photons needed, although at this time they lack the necessary repetition rate. Nevertheless, it is reasonable to assume that appropriate lasers will be available by the time they are needed. In this way one can generate collisions between gamma rays, or between gamma rays and electrons. It was V. Telnov (see sidebar for Telnov) who first realized the possibility of gamma-gamma colliders. Subsequently many important aspects have been studied by him and by other workers.

X.4 The Compact Linear Collider (CLIC)

Meanwhile, and in parallel, CERN will pursue its own alternative 30 GHz technology to prepare CLIC for the day when more than 500 GeV (and perhaps more than 1000) GeV per beam is required. Superconducting breakdown will limit the superconducting cavities of the ILC to less than 60 MeV/m, while CLIC holds the promise of 150 MeV/m. Although CLIC already seems to be somewhat more cost effective than superconducting technology, its R&D has some years to go before a definite proposal can be made. What in the past might have been the two rival projects of different countries and continents has been resolved, very rationally, into a global collaboration suiting the needs of a world community as well as the particular aspirations of its members. ILC is the only technology that is ripe, and it is appropriate to give it most of the available resources so that it will ready in time; while CLIC needs to use the resources of CERN, limited until LHC is over, to aim for a later ripening and a harvest when the even higher energies that are possible with CLIC are needed.

In the last section we hinted that higher accelerating gradients than are contemplated by the ILC are needed to build a linear collider of 1 TeV per beam or higher within a reasonable strip of real estate. Such gradients are only possible at higher frequencies. The value of 30 GHz has been selected by the CLIC group. At this frequency the RF power source is the crucial question. You cannot buy a tube that will power a 30 GHz linac off the shelf, and to design one is still a challenge. In a klystron, power is taken from an electron beam, modulated at the required frequency, passing through a cavity. The microwave power is transferred to the linac by a wave-guide. In the CLIC scheme, the main linac receives its power not from klystrons, but from an intense electron beam in a separate linac, running parallel with the main linac. The beam in the driver linac is accelerated to a modest energy by cavities working at the LEP RF frequency, where klystrons are readily available. Like the klystron beam, the CLIC drive beam must be modulated at the required frequency of 30 GHz. The linac beam consists of intense electron bunches, only a few millimeters long: less than one wavelength at 30 GHz. Collector cavities and transfer structures tap the power and pipe it over to the main linac. The R&D for CLIC started at CERN under Wolfgang Schnell and later Kurt Hübner, Jean-Pierre Delahaye and Ian Wilson (see sidebars for Schnell and Hübner) in the 1980s and has grown in recent years into a collaboration of institutions in England, Finland, Germany, Italy, the United States, Russia, Japan, France, and Sweden.

The beam in the main linac is prepared, like that of the ILC, from electron and positron sources by a sequence of damping rings, bunch compression, and special rings for compressing and combining bunches. In particular self-field effects are more severe at higher frequencies, where the aperture of the linac cavities is small and must be carefully handled. But, because the frequency is high, the accelerating gradient in CLIC is more than four times that of the ILC (150 MeV/m vs 35 MeV/m).

The first R&D problem was to design "transfer structures" that would take power from the drive beam and transfer it to the main beam. The second was to produce a very intense drive beam with bunches only 1 cm apart (appropriate for 30 GHz). These basic problems have been solved, although some only "on paper" and experimental demonstration is now required. The drive beam is made by starting with a beam accelerated

Valery Ivanovich Telnov (1950–)

Russian physicist
Young Investigator Award, Budker Institute of Nuclear Physics, Russia
Special Award from the Stanford Linear Accelerator Center, USA
Soros Science Education Foundation Award
Fellowship of the Ministry of Science and Education, Japan

Telnov was born in a small village, Podgornoe, in Siberia. His father had been blinded during the Second World War. His mother was editor of the district newspaper. While still in 7th grade he entered the All-Siberian Physical-Mathematical Olympiad and won the physics prize. He completed his high school education at the Special Physical-Mathematical boarding school at Novosibirsk University. At the Novosibirsk State University he received an MS (1972) and a PhD (1973) while doing research at the Budker Institute of Nuclear Physics. He has always been associated with both institutions and is now both a Chief Scientist at the Budker Institute and a full Professor at the University.

Although blessed with considerable theoretical ability, Telnov became an experimentalist in particle physics. He has many accomplishments as a particle physicist, but here we emphasize his very important contribution to accelerators. In 1980, during a workshop, he had the idea of using an electron linear collider not to collide the electrons themselves, but to use them first to produce energetic photons and then study their collisions (as is described in Section X.3). Later, after the workshop and together with I. Ginzburg, G. Kotkin, and V. Serbo, he developed the concept of a photon collider based on Compton scattering — a concept that could only occur to a mind well versed in accelerators and both experimental and theoretical particle physics.

The first article on this idea was rejected twice from Soviet journal *Pisma JETPh*, and then also from *Physics Letters*, but an extensive paper on the photon collider was eventually published in *Nuclear Instruments and Methods* and is now the most cited paper from that journal!

Telnov has traveled widely outside Russia, at SLAC (California), CERN (Switzerland), KEK (Japan), and at DESY (Germany). He was married in 1973 and he and his wife have a son who obtained a PhD in physics from the University of California in Berkeley. His hobbies, besides physics, are skiing, travel, diving, and photography, and he also finds time to think about the mysteries of the universe.

to 2 GeV in a structure and then sent through compressor and combiner rings to change the 3 GHz intra-bunch spacing of 10 cm down to 1 cm. This tricky process is to be experimentally tried in the CLIC Test Facility 3 (CTF3), scheduled for completion by 2007 (about the time LHC will come on stream). Other items will also be tested at that time. For example the high gradient of 150 MeV/m could not be obtained in copper

Wolfgang Schnell (1929–)

German physicist
CERN's radio-frequency expert
Inventor of CLIC

Wolfgang Schnell was the son of a medical doctor and was born and brought up in Kaiserslautern during Nazi times in Germany. He was spared military service as the War ended just a couple of weeks before he was 16. He went to Heidelberg University but did not wait to receive a PhD. The prospect of joining the young team of international experts recruited by CERN in its very early days was too attractive to miss.

Working under the leader of the Radio Frequency System Group Leader C. Schmelzer, his job was to find a way of making sure that the radio frequency driving the accelerator cavities of the PS remained in step with the bunches of particles as they circulated in the ring. This would ensure that the beam remained in the centre of the doughnut-shaped vacuum chamber. There were two alternative proposals, one involving the control of the accelerating voltage and the other, its phase. It was a youthful Schnell who pushed through the latter strategy which proved much more effective.

The CERN PS and the AGS in Brookhaven faced a new problem of making this system work as it accelerated up to "transition" where a swift shift in phase of 180 degrees would be needed to keep control. It was largely thanks to Schnell, working away busily in the centre of the PS ring that the first accelerated beam in that machine began to be accelerated from injection up to the dreaded transition energy. Taken by surprise at the speed of this progress he had to, that same evening, rapidly improvise a switch for the phase lock, to be triggered at the point of transition. He built this circuit in a Nescafe tin. This modest but crucial device was to immediately open the door to full energy acceleration and to prompt the world's congratulations on CERN's success in bringing the first big alternating gradient proton synchrotron to life.

Schnell's next challenge was to make the ISR's beam stacking scheme work. Here again he was breaking new ground and as he moved on to become responsible for that machine's entire performance. He shepherded his team through a number of baffling obstacles with names like "the brick wall" and "pressure bumps," which had to be circumvented to raise the ISR's current from less than an A to more than 30 A — an intensity that could be stored for a day or more. It was during this period that he helped invent a "stochastic" beam measurement method which was later to prompt Simon van der Meer to propose the method of cooling antiprotons for the Nobel prizewinning SPS collider (see sidebars on van der Meer and stochastic cooling in Chapter VI).

Next of course, came LEP. Together with E. Keil and C. Zilverschoon he worked out the design study and played a crucial role in winning the project for CERN. Summoned to Hamburg to a meeting where he was supposed to listen quietly to DESY's bid for the new machine he took advantage of the fact that the slides for their presentation could not be shown and slipped in his own exposition of CERN's plans.

Very much the same scenario repeated itself as Schnell, with LEP behind him, proposed the CERN Linear Collider at a time that DESY had its own plans for other kinds of linear colliders.

Wolfgang Schnell pursues his ideas with a passion and single-minded commitment that is unsurpassed by others in the field. His painstaking analysis of how best to power a 30 GHz linac with a second high current linac alongside it was refined over a period of several international accelerator conferences to become the Compact Linear Collider, which is CERN's bid in this field as we write. He epitomizes all that is good in the German character — a high degree of proficiency in his field with a dogged pursuit of excellence and all of this with a healthy sense of humor and respect for his colleagues.

In his later years he has spent much of his time pursuing quite a different field of theory, trying, unsuccessfully so far, to persuade the highly academic world of theoretical physics to take note of his theory which explains the universe, predictably perhaps, in terms of the resonant modes of a cavity. To everyone's surprise this generates the precise relationship of the masses of a few score of fundamental particles. Time will tell whether Schnell or Higgs wins in the end!

---

cavities (which degraded under the intense field), but seemed to be feasible in molybdenum or tungsten cavities. Already the high gradient of 150 MeV/m has been achieved for short pulses and without significant damage to the cavities. CTF3 will attempt to reproduce this performance with pulses of the correct length.

Finally, we want to note that there is a modest program at SLAC investigating accelerators operating at the very high frequency of 90 GHz while CLIC has just lowered its optimum frequency to 12 GHz.

X.5 Spallation Neutron Sources

If we look away from high energy particle physics, there are many other research fields which use accelerators of relatively low energy — a GeV or so — but pushed to the highest possible beam intensity. A typical case involves accelerators used to produce neutrons by spallation.

The use of accelerators as spallation sources of neutrons has been pioneered by a number of laboratories, such as Argonne and Los Alamos in the US, Rutherford-Appleton in England, and the Paul Schering Institute (PSI) in Switzerland. In a spallation source, protons are typically accelerated to 1 GeV in a rapid cycling synchrotron or linac followed by a pulse-compression accumulator ring. The pulse is then fired at a (cooled) heavy metal target, and each proton produces 20–30 neutrons via primary or secondary reactions. The neutrons are then slowed (cooled) down to energies typically in the milli-electron-volt (meV) range by means of a "moderator": a material made of light atoms such as water or liquid hydrogen.

At these energies, the wavelength of a neutron (remember, particles can be seen to act as waves under the right conditions!) is about the same as x-rays, and they can scatter into diffraction patterns much as those seen in synchrotron-radiation sources. However, neutrons scatter off of the nuclei of atoms in targets, not the electron clouds, so they turn out to be much more sensitive to light elements than are synchrotron-radiation sources. Thus neutron diffraction can see hydrogen and deuterium (and can tell the difference between them) in the presence of heavier elements, which is just not possible with x-ray sources. In this way, neutron-based spallation sources are complementary to synchrotron-radiation sources; they both serve the same communities of material scientists.

Reactors of course are copious sources of neutrons, and the field of neutron diffraction began in the 1940s at research reactors. The value of this technique has been recognized with the Nobel Prize awarded to Shull and Brockhouse in 1994. As the wavelength of the neutron (hence its velocity) are all-important, selecting neutrons of the correct velocity, or tagging neutrons according to their velocity, is critical. In reactors, where neutrons of many velocities emerge from a port continuously, selecting velocities is very difficult and the efficiency of neutron use is very low. With accelerators, on the other hand, a short pulse (of about a microsecond or less) hitting a target produces neutrons that all leave the moderator at about the same time. As the speed of these very low-energy neutrons is only a few 100 meters per second, it takes many milliseconds for the neutrons to travel to the target station, typically a few 10's of meters away. (It is interesting to note that the effect of gravity must be taken into account; the neutrons can fall several centimeters along their flight path!) By measuring the time the neutron takes in going from the moderator (the "start" pulse) to the detector after scattering from the sample (the "stop" pulse), one can tag every neutron with a distinct wavelength, so the net efficiency of neutron use is much higher than with reactor-produced neutrons. In fact, the productivity of a 1 MW spallation source is viewed as comparable to that of a 60 MW research reactor.

Figure 10.5 gives an example of the complex structures that can be mapped with neutron beams. In addition, neutron diffraction is used for studying high-temperature superconductivity, magnetic properties of

Kurt Hübner (1937–)

Austrian physicist
Director of Accelerators at CERN, 1994–2001
European Physical Society Accelerator Prize

Kurt Hübner studied mechanical engineering in Austria, for which he received a master's diploma. While working parttime in a firm of the family he was encouraged by a professor of experimental physics to start again as a student, which he did. He attended courses in physics, made all the exams required for the master's, and, finally, defended his thesis.

Once his education in Austria was completed, in 1964, he joined CERN as a fellow. He remained at CERN for his full career. His very first work was on the CESAR electron storage ring project in preparation for the Intersecting Storage Ring (ISR) project. He then went on to join the team responsible for the design and construction of ISR.

At the end of the 1970s, once the ISR was built, he played an important role in the design studies for LEP, in particular proposing that the existing proton synchrotrons — PS Booster, PS and SPS — might be adapted to accelerate electrons and positrons, and be used as the injectors for the new collider. This reuse of existing resources was important in convincing CERN's member states to approve the project. In 1983, when LEP had been approved he was given the job of coordinating the implementation of this scheme. In addition to his involvement in the commissioning of LEP, he took part in drawing up the proposals for the LHC and in the studies for CLIC, the project for a future linear collider, which brought CLIC from an interesting theoretical concept towards practical reality.

Modest and easy to work with, he has always demonstrated a painstaking analytic approach to technical work, examining all the options and discussing them with colleagues at length to arrive at a final and logical result. In this he often draws on his extensive memory of all previous relevant work. Once the decision is made, he takes pains to defend it against all opposition. This is exactly the approach he employed within the respective teams which got the basic experimental feasibility tests for CLIC done, implemented the upgrade of LEP in energy, and initiated the CERN Neutrino Beam to Gran Sasso (CNGS).

He became leader of the CERN's Proton Synchrotron Division in 1991, before being appointed its Director of Accelerators in 1994 — a post, which he held until 2001. Since then, he has represented CERN in the Organisation for Economic Cooperation and Development and International Committee for Future Accelerators working groups and has also been a member of the committees of many international accelerators, including the scientific councils of DESY and Laboratoire de l'Accelerateur Linéaire (Orsay).

In 2002 he was awarded the European Physical Society's Accelerator Prize and whose citation read: "He provided guidance for generations of accelerator physicists and engineers, thereby contributing immensely to the prosperity of accelerators at CERN and many other laboratories around the world."

Fig. 10.5 A picture taken at the British neutron facility ISIS showing a hybrid microporous organic-inorganic solid. Neutron diffraction is particular in its sensitivity to light elements such as hydrogen, deuterium, carbon, nitrogen and oxygen, and thus provides an ideal tool for structural studies of such materials. Synchrotron radiation, on the other hand, is sensitive to heavy elements, so the two approaches are complementary.

materials, quality of welds and strains in critical structures such as aircraft wings.

Alternative sources of neutrons are reactors but, unlike neutrons from accelerators, the reactor neutrons are not pulsed. The pulses of a spallation source allow time of flight techniques to be employed to reveal the transient response of the target materials. At present there is considerable "user pressure" to build ever larger spallation sources.

High beam current is needed, and every trick in the accelerator builder's repertoire must be used to ensure that the beam does not break into wild instabilities due to its electrical image in the walls of the vacuum chamber as well as in every component through which it passes. Losses lead to radioactivity, but also, loss of even a tiny fraction of the beam during acceleration might easily burn a hole in the beam pipe; and even when the beam has been extracted the danger is not passed. There is a powerful thermal shock as megawatts of beam power impinge on less than a kilogram of metal that forms the spallation target.

Two spallation sources, a synchrotron called ISIS at the Rutherford-Appleton Laboratory in the UK and a cyclotron driven spallation source at PSI in Switzerland, share the honors for the most powerful sources in operation. An even larger spallation source (the SNS) has just been completed in the US, at Oak Ridge, while serious consideration of a similar spallation source has been going on for many years in Europe. The Japanese are constructing a large accelerator complex at the JAERI, Tokay site, the major part of which is a spallation source. In 2005 the Chinese initiated a design study for their own spallation source (the CSNS).

The SNS in the US is typical of this new generation of sources (see Fig. 10.6). When run in up to its design specification it will produce an average proton beam current of 1.4 mA and at 1 GeV will deposit 1.4 MW of power in the target and its surroundings. Each proton pulse on target is 0.7 microseconds long, short enough to provide excellent wavelength tagging of each neutron, with enough intensity to produce excellent images of the most complex material strutures. The injector for the synchrotron accumulator ring embodies all the latest technology. An ion source is followed by an RFQ, then an Alvarez linac of 7.5 m followed in turn by a superconducting linac of 331 m. More than 1000 turns of beam are fed into the accumulator ring, 248 m in circumference and then shot at a mercury target.

The cost of the SNS — about 1.5 B$ — is comparable with the big machines of particle physics. Of the many aspects of accelerator physics and engineering involved in constructing and operating such a major facility, the complexity of remote handling for maintenance and the suppression of uncontrolled beam loss

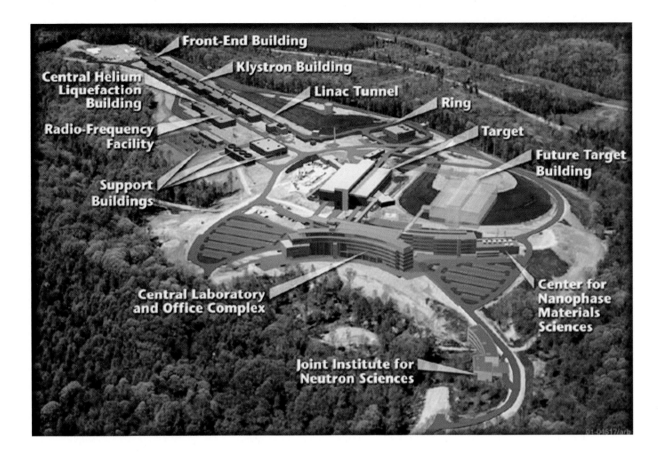

Fig. 10.6 An overview of the Spallation Neutron Source (SNS) site at Oak Ridge National Laboratory, showing the various components of the facility.

are perhaps the most demanding. If the complex is not to become too hot (radioactively) to handle, losses must not be more than a watt per meter length.

It comes as a surprise to find that the SNS is the first US accelerator to be a joint effort of several of the major national laboratories under the Department of Energy. Berkeley produced the ion source and the RFQ, Los Alamos produced the normal-conducting Alvarez and side-coupled linacs, Jefferson the superconducting linac, Brookhaven the accumulator ring, Oak Ridge the target, and Argonne coordination of the neutron-scattering instrumentation.

The high beam intensity and target region design problems of the spallation source are shared by other projects of the future, which will be discussed in the remainder of this chapter. These include neutrino factories, where the secondary particles from the target, muons, decay to produce a beam of neutrinos directed through the earth to some distant location (see Fig. 10.7); muon colliders, which exploit the fact that, in the energy range of interest, muons are like electrons without the embarrassment of synchrotron radiation; and proton drivers to stimulate fission in fail-safe sub-critical nuclear reactors that could use thorium

as a fuel and burn up nuclear waste. All these exciting projects require intense proton beams (in the range of a few GeV) directed on an external target.

X.6 Rare Isotope Accelerators

The last 80 years of nuclear physics has mainly been devoted to studies of nuclear properties using accelerated beams of stable isotopes. The result is a detailed understanding of nuclear behavior which has had practical application in the fields of nuclear medicine, nuclear power and weaponry. The next major project that nuclear physicists plan is an accelerator of radioactive species of very short half-life — the so-called non-equilibrium nuclei.

Beams of these rare species have already been produced and studied at the ISOLDE facility at CERN, the Bevalac at LBL, and at the superconducting cyclotron at Michigan State University, but now new facilities of expanded capability are called for. There are four projects of this type. The first, already completed, is the Radioactive Isotope Beam Facility (RIBF) at RIKEN, Wako, in Japan; the second, under construction, is the Facility for Antiproton and Ion Research (FAIR) at GSI,

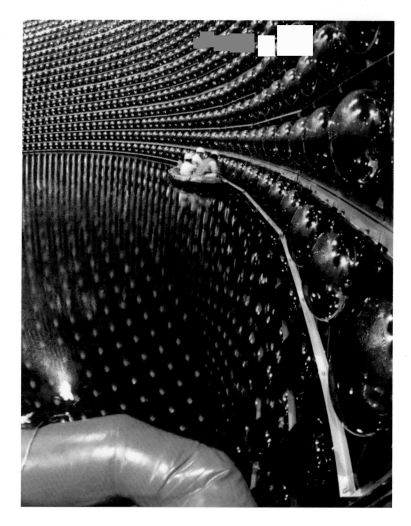

Fig. 10.7 An underground detector called Kamiokande, near Kamioka in the mountains of Japan, is one of several neutrino detectors in the world. These particles are often described as "ghostly" — they carry almost no mass and hardly interact at all with matter — so these facilities are made up of thousands of tons of matter, monitored in darkness by detector elements that can pick up the slightest scintillation. (The workers in the rubber raft, servicing photomultiplier tubes, give an idea of the scale.) Kamiokande began taking data in 1996 and played a key role in solving the "solar neutrino problem" — the great gap between the theoretically predicted rate at which the Sun emits neutrinos and the flux actually observed on Earth. As it turns out, neutrinos oscillate between three types or "flavors" in the Earth's upper atmosphere, so there is no flaw in either the Sun itself or the otherwise very successful theories of how stars operate. The solar neutrinos, for the most part, were just changing to a previously unexpected kind — which, unsurprisingly, previous detectors were not designed to see. An implication is that the rest mass of the neutrino is tiny but finite; this has huge implications for the fate of the universe, since there are so many neutrinos that even their near-infinitesimal individual mass adds up to a large figure. Kamiokande had a severe mechanical failure in 2003 — no one was hurt, but hundreds of these photomultiplier tubes fell and shattered. It was put back into service, and soon an upgrade called Super-Kamiokande will be completed. Some of these facilities receive neutrino beams from accelerators; the distance between the two is calculated to allow for the desired "flavor mixing" while minimizing interference.

Darmstadt in Germany; the third is SPIRAL2 at GANIL in Caen, France; and the fourth is the Rare Isotope Accelerator under consideration in the US.

The Japanese facility consists of adding two new conventional cyclotrons and a superconducting cyclotron to an existing linac and cyclotron. (The superconducting cyclotron shown in Fig. 10.8 is the largest cyclotron in the world.) In this way they will be able to accelerate any ions created by the earlier cyclotrons. The facility is impressive in size and, being already completed, provides the Japanese with a rare ion accelerator many years before other facilities are brought on line.

The Germans at Darmstadt, at the Gesellschaft für Schwerionenforschung (GSI), have initiated a major expansion of their facility. The new project, called Facility for Antiproton and Ion Research (FAIR), was initiated in February 2003, with the German Government agreeing to support 75% of the cost and the remainder to be contributed by European International partners. The facility will have at its heart a double ring superconducting synchrotron facility with a circumference of 1,200 m. There will be a system of cooler-storage rings with electron and stochastic cooling

at various stages. The present accelerators at GSI will be used as injectors. The expectation is to have the project completed in 2010.

The facility will be used to study nuclear matter under various conditions, including those that existed just after the Big Bang. Besides nuclear matter physics and rare isotope studies (much as seen for RIA), they will study atomic physics, plasma physics, and the physics of antiprotons.

The French, in their facility, are constructing a linac that will produce exotic isotopes both by in-flight fragmentation and by smashing light nuclei into targets of heavy nuclei to break off pieces (as is already done at CERN), which can then be reaccelerated for reaction studies with these exotic projectiles. They expect to have their facility ready in 2011.

Both Michigan State University and the Argonne National Laboratory are developing designs for the Rare Ion Accelerator (RIA) project. The US Government had halted the project in early 2006, but then later in the same year a report strongly suggested that it should proceed. Currently the Department of Energy (2007) is studying just how much of the project it will support. In

Fig. 10.8 The large superconducting cyclotron, the last in a chain of four cyclotrons located at the Nishina Laboratory at RIKEN in Wako, Japan.

its entirety, the RIA will produce high quality beams of isotopes at all energies. It is based on a 400 kW superconducting heavy-ion driver linac, which produces a radioactive species to be quickly accelerated in a secondary superconducting linac.

The primary accelerator will provide beams of all elements from protons at 900 MeV to uranium at 400 MeV per nucleon. Its production chain will consist of a flowing liquid lithium target followed by extensive fragment-analysis systems and degraders, with a high-purity gas cell to finally thermalize and stop the separated fragmentation products. Once the various isotopes are swept out of the stopping area, they are collected, ionized, separated in a high-resolution isotope separator, fed to a low-frequency RFQ.

X.7 Neutrino Super Beams, Neutrino Factories and Muon Colliders

Recent years have brought a strong interest in neutrino physics. The "solar problem" — the fact that the flux of neutrinos from the sun is a factor of two low — has only recently been "solved." Experiments have shown

that the neutrino has mass and that the neutrino from the sun — emitted in beta decay — can mutate into a second type of neutrino, which is emitted in muon decay. The detectors that were only sensitive to one type of neutrino registered a smaller signal than would have been expected without neutrino conversion.

The realization that neutrinos are able to change from one kind to another implies that they have mass, something that cannot be explained by the Standard Model and which gives high-energy physicists their first scent of "big game" for many a year. Neutrinos oscillate from one type to another over huge distances, and to study this, the Japanese have sent an accelerator-produced beam of neutrinos to a detector hundreds of kilometers away. (Neutrinos interact only weakly with matter and only few of them are lost even in passing through the center of the earth.) Similar programs are now underway at Fermilab (Illinois to Minnesota) and at CERN (Switzerland to Italy).

These experiments use neutrinos produced by the decay of particles produced by proton beams from existing synchrotrons, but people are thinking of "super beams" from even more powerful accelerators. Another method is to use "beta beams" of accelerated radioactive species that decay in flight to directly produce neutrinos from their beta decay. The ultimate neutrino source is the "neutrino factory": a storage ring where muons decay into a beam of neutrinos. This idea stems from an international collaboration of physicists exploring the acceleration of muons to make a storage ring for a muon-muon collider, an effort led by Robert Palmer (see sidebar for Palmer).

Although muon colliders share a common technology with the neutrino factory, their physics motivation is quite different. We explained above that linear colliders seem to be the only imminent way to extend the energy range of circular electron-positron colliders, but there is the alternative of building a circular collider for muons and their antiparticles. Muons are leptons, like electrons, and share with electrons the advantage of an unambiguous point-like structure. Physics experiments with muons are very similar to those with electrons of the same energy, but because the muon is 200 times heavier than the electron, it will not radiate significant amounts of synchrotron light even at the energies at which the present generation of proton colliders operate.

However, there are two major difficulties that must be overcome in accelerating muons. The first is that the muon is not a stable particle like the proton or electron, but decays in about a microsecond — barely time to make a turn in a large storage ring. Fortunately, as a consequence of the special theory of relativity, the time and distance it travels before it decays increases in proportion to the energy. This means that, provided the storage ring operates at about 100 GeV or more, the colliding beams will exist for about one millisecond in the laboratory reference frame: long enough to see some events.

A second major problem has to do with the production of muons. The only way we know how to produce enough muons is to take an intense beam of protons and fire them at a target. Pi mesons will be produced, and they almost immediately decay into muons. Unfortunately the spread in angles of the muons is far larger than the acceptance of most accelerators (an exception is discussed below) and hence the muons must be "cooled." However, all the usual methods of cooling are too slow when compared to the muon lifetime; the only possible method is ionization cooling.

The concept of ionization cooling (see sidebar on cooling) was first proposed by Budker and Skrinsky and further developed by others (see sidebar for Skrinsky). In the overall process of losing energy (momentum) by

Robert B. Palmer (1934–)

American physicist
Panofsky Prize of the American Physical Society, 1993
American Physical Society R.R. Wilson Prize, 1999

Robert Palmer was born in Harrow, near London, and attended boarding schools in Essex and in Yorkshire. At Imperial College he was an undergraduate, then a graduate student obtaining a PhD in 1959. In 1960 he went to Brookhaven National Laboratory where, as a participant in the Bubble Chamber Group, in 1964 he was co-discoverer of the Omega Minus. Later, (1971–1972) he went on sabbatical leave from Brookhaven, to CERN.

Over the years he made many important contributions to high energy and accelerator physics. He designed the optics, hydraulic expansion systems and superconducting magnets for bubble chambers. He proposed Inverse Free Electron Acceleration and a method of Momentum Stochastic Cooling commonly used in many accelerators. He was instrumental in developing superconducting magnets, including the first "two-in-one magnet," the configuration used in the 27-km circumference Large Hadron Collider.

Apart from these three very different and original ideas, he has also contributed to administration of physics in ways too numerous to mention here. He founded the Accelerator Test Facility at Brookhaven, he was Associate Director for High Energy Physics at Brookhaven (1983–1986), he was a member of the High Energy Advisor Panel of the Department of Energy (1983–1986), and he had a joint appointment at the Super Conducting Super Collider (1990–1999), and was on their Science Policy Committee (1989–1991). He played a crucial role in initiating the Neutrino and Muon Collider Collaboration as well as making very many contributions to the science of these devices.

Bob Palmer is a physicist with a wide range of interests and unusual ability. His contributions to the science have been remarkably significant and varied. His hobbies are sailing, skiing, climbing and travel and he has many accomplishments in these fields. With his wife he has founded "Palmer Photo Cards," which combines his talent in photography with business sense. He has had a joint appointment at SLAC (1987–1996), and been associated with the Lawrence Berkeley National Laboratory since 1997.

ionizing the atoms of an absorber — and then being reaccelerated in the forward direction — particles take up new trajectories closer to the axis of the beam. After many such operations they can be made to fit the acceptance of a storage ring. On paper this looks possible, and a large international cooling experiment, the Muon Ionization Cooling Experiment (MICE), is under development at the Rutherford-Appleton Laboratory in England to prove it. (See Fig. 10.9.)

Beam Cooling

To stand a chance of producing a rare particle among the many events which occur when the beams collide, many particles must be crowded into a very narrow beam. (See sidebar on Luminosity in Chapter VI.) In Chapter VI.2 we describe how difficult it is for the synchrotron rings, which prepare the beam for a collider, to make a beam that is dense enough. One may lay down beams side by side in a synchrotron, but each will occupy a different band of transverse or longitudinal energies, rather like streams of amateur drivers swerving from side to side in their lanes on a motorway. A theorem developed in the early 1800s by Liouville states that as long as the individual cars can be thought of as a continuous stream of traffic and as long as the motion is energy-conserving (technically "Hamiltonian"), the traffic cannot be merged; i.e., the beams of particles cannot be superimposed. All one can do is to "raise driving standards" to reduce the width of each band of energy and squeeze the streams closer.

However, it is possible to circumvent Liouville's theorem by considering the individual particles or by having non-energy conserving processes. The resulting increase in particle (technically "phase space") density is called "cooling." Thus cooling is a process which reduces the energy of a particle oscillation in either one of three possible directions — horizontal, vertical, or in the direction of acceleration — to bring the width of the spread in energy among the particles within the band and bring the bands closer. It is called "cooling" because we can express the kinetic energy of these oscillations, like the energy of molecules in a gas, by an equivalent temperature — the hotter the gas, the more energetic is the motion.

A simple, but very effective, cooling mechanism is the radiation of photons from a circulating beam of electrons. (The beam must be accelerated — in this case by moving about a circle — in order for it to radiate.) The motion of the electrons alone is NOT energy-conserving. The complete system of electrons and photons IS energy conserving, and Liouville's theorem applies, but if one looks only at the electrons — we don't care about the photons — then energy is not conserved and, in fact, the electrons can be made to cool. Radiation cooling in special "damping rings" is at the heart of linear colliders.

In another sidebar in Chapter VI we describe stochastic cooling — perhaps the most famous method of cooling because it was used for the Nobel prize-winning discovery of the W and Z. It works by using the granular nature of a beam of particles, i.e., that the particles are not really a continuous stream (but almost so). Unlike radiation cooling, this method works also for heavy particles, like protons.

There are a number of other ingenious methods of cooling. As with the above methods, each is either special, or particularly effective, for particular particles at particular energies. One of these — electron cooling — was first developed in 1967 at Novosibirsk under the aegis of Budker (see sidebar for Budker in Chapter VI). A beam of heavy particles, antiprotons or ions for example, travels along the same path as a beam of electrons with the same velocity, sharing its surplus energy with the electrons. There is a principle of physics called "equipartition of energy" that equalizes the kinetic energy of the two species of particles through the forces of their mutual electromagnetic fields. After this equipartition is over, the anti-protons, being almost 2000 times heavier than the electrons, will only retain about one-fiftieth of the velocity spread of their lighter companions and, if the heavier particle originally had the lion's share of the energy, it will be "cooled" by this factor of 50. At high intensity of proton or ion beams this kind of cooling is a much faster process than stochastic cooling, and is particularly effective when the proton, ion or antiproton beam is circulating at low energy. At first it was considered as a rival candidate to the stochastic method of cooling antiprotons for use in colliders, but stochastic cooling turned out to be much more useful at higher energies, large emittances and for a relatively small number of antiprotons. Recently electron cooling has been employed at the high-energy anti-proton recycler at Fermilab and it is being developed for high-energy ions at RHIC (Brookhaven).

Laser cooling is another method and is also a fascinating practical application of modern physics. The idea is applied to a beam of ions circulating in a storage ring. The laser excites the ion from one direction, and the ion then radiates in an isotropic manner and so cools. (The violation of assumptions for the Liouville theorem is very similar to that in radiation cooling.) However the radiation must be exactly to the initial state, as the process must be repeated a million times to cool any one ion.

Ions speeding towards the laser beam see the laser beam shifted in frequency and hence in photon energy — the Doppler effect. An ion of a particular and precise velocity can see photons whose energy exactly matches the difference in its electron energy levels. When such a resonance occurs, energy is transferred between the photon and the ion, speeding it up or slowing it down. The trick is to sweep the laser frequency, or the photon energy, backwards and then forwards into the distribution of velocities in the beam, squeezing it from both sides into a much narrower band. An alternative is to sweep the particle beam energy — often done with an induction unit.

The separation of the quantum energy levels of the electrons orbiting the ions must be the same as the energy of the laser photons; and, most importantly, have only a ground state and one excited state connected by the laser photon. Only a few ions have these characteristics, but they have been cooled most effectively by this method.

Yet another method is used for the beams for muon colliders. Muons offer a route to extend circular lepton accelerating storage rings to energies of several TeV, but the acceleration and storage must occur in a few milliseconds to compete with their rapid decay. Like antiprotons, the muons must be cooled. Because of the time constraints, single-pass cooling or perhaps cooling over a few passes, is the only solution. In the so-called "ionization cooling" method, particles pass through an energy absorbing plate and lose momentum in the direction of their trajectory to the electrons of the absorber material. (Thus the assumption of energy conservation, in Liouville's theorem is violated.) If they are "hot" and therefore moving at a small angle to the axis of the beam, some of the transverse as well as longitudinal momenta will be lost. An RF cavity following the absorber replaces the longitudinal but not the transverse momentum. Repeated a few times, the process will cause a steady reduction of transverse momentum — reducing the angle of the path and cooling the beam. Longitudinal cooling can also be accomplished; it requires bending magnets. This approach is being actively studied at this time.

Alexander Skrinsky (1936–)

Director of the Budker Institute of Nuclear Physics
Fellow of the American Physical Society
Member of the Royal Swedish Academy of Science
American Physical Society R.R. Wilson Prize, 2002
Academician of the Russian Academy of Sciences

"Sasha" Skrinsky, as he is known to the world of particle accelerators, was born in Orenburg on the Ural River exactly between Europe and Asia. His father was an automotive engineer and he has one younger brother who is an accomplished car production engineer and administrator. He studied in many different grammar schools, due to WWII, as his father was in the army and they moved a great deal. Finally, after even some time in Berlin, he graduated from high school in Gorky (now reverted to its old name of Nizhni Novograd). He became interested in physics at the age of nine and although his high-school physics teacher was "absolutely very bad," he had a fine mathematics teacher and went on to study physics at the Moscow State University, where he graduated in 1959.

Skrinsky has always been at the focus of storage ring development at Novosibirsk in Siberia since he and Budker (see sidebar for Budker in Chapter VI) built the first electron-electron collider VEP-1 — a revolutionary machine, constructed in parallel with the first Princeton-Stanford collider in the US.

He and his team went on to build many other colliders at the Budker Institute, Novosibirsk (see sidebar in Chapter VI on Novosibirsk). These are VEPP-2, VEPP-2M, followed by VEPP-4. For this and much of his subsequent work his team received a series of USSR, and later, Russian, and US medals and prizes. All these awards were for an imaginative team — probably the best there has ever been in accelerator physics and which relied upon Skrinsky for its leadership and inspiration.

The intellectual fertility of the team was already legendary at the time of Budker's death in 1977. Skrinsky, taking over the leadership of the Lab, and his colleagues added to this reputation with a steady stream of innovation. One of these was a method of obtaining the longitudinally polarized beams for colliding beams in storage rings, later to be used for the electron ring of the HERA collider in Hamburg (see Chapter VI). The method is also used in the project of the Novosibirsk Charm/Tau Factory.

Spin resonance in storage rings and their (multiple) crossing had been studied both theoretically and experimentally (1971), and the ways to excite resonance by the external RF field were suggested. On this basis, a method was suggested for high precision measurements of the elementary particle masses by resonance depolarization of the electron-positron colliding beams. Developed and realized mainly in Novosibirsk, this method allowed establishment of a precise mass scale up to the level of 100 proton masses with the accuracy of 30 parts per million.

Skrinsky made other fundamental contributions to the discovery of a method of electron cooling that had been suggested by Budker in 1965 and realized in 1975. In 1978, he was among the first to present, together with Balakin and Budker, the conceptual project of the linear collider, leading now, decades later, to today's most adventurous project: the International Linear Collider (ILC).

Looking even further into the future, work on finding the way to muon colliders of hundreds of GeV and higher, started early at INP by Budker, Skrinsky and their colleagues. This led in the 1960s to the important invention of the ionization cooling method.

Together with G.N. Kulipanov and N.A. Vinokurov, Skrinsky made a substantial contribution to the development of synchrotron radiation generators including, with the help of, linear (or several turns) accelerators with energy recuperation. In the field of free electron lasers Vinokurov and Skrinsky suggested a very important modification based on electron storage rings — the optical klystron. Such a laser at a mean power of up to 100 kW (with an energy of electrons over 50 MeV) is under construction in Novosibirsk.

Seen from the West, throughout the difficult days of the "Cold War" and the happier times that followed, Sasha Skrinsky has remained a true friend of accelerators and those that build them, pursuing his agenda of innovation with ceaseless determination. To colleagues in the West he always shows a smile not merely of recognition but of true friendship, and to those in the East he must seem to possess an unparalleled intellectual resolution focused to generate new ideas.

A neutrino factory will have a proton driver, typically with a beam power in the MW range; a liquid metal target; a capture channel; a cooling channel; an accelerator; and finally a storage ring in which the muons circulate for about 500 turns, as they decay, to create a neutrino beam. A muon collider will be rather similar but the cooling necessary to make a sensible number of collisions is much greater than for a neutrino factory.

The study of new and exciting ideas like neutrino factories is fertile ground for alternative technologies to grow and overtake the latest ideas; almost before they have been tried. Ionization cooling may well prove superfluous if the new FFAG accelerators which the Japanese have been building can be used. These machines have such a large acceptance that they can collect enough muons for a neutrino factory without the need for cooling. Of course ionization cooling and MICE will still be relevant for a muon collider.

For both a neutrino factory and a muon collider, new muon beams will have to be produced, accelerated and cooled to refill the storage ring several times per second. The acceleration and cooling processes must be very fast and efficient because until the muons get up to speed a lot of them will decay. Most schemes include a recirculating linac (see Chapter III) or an FFAG (see

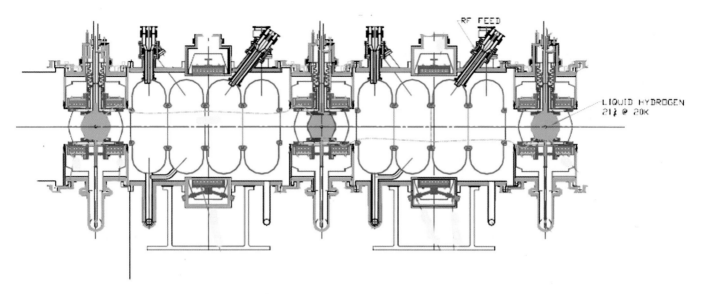

Fig. 10.9 A diagram of the muon cooling experiment MICE being carried out at the Rutherford Appleton Laboratory in England.

Chapter II), as well as ionization cooling. The technology necessary for a muon collider is in its infancy and probably will be studied by first undertaking a less demanding task; namely a neutrino factory.

X.8 Accelerators for Heavy Ion Fusion and for Creating High Energy Density Plasmas

The subject of energy is, of course, of great concern. A small discussion of long-term sources of energy is presented in the sidebar on Energy. There are many books on the subject of energy, but a brief description of long-term sources is germane to our book, as accelerators have a potential role to play.

There are two possible methods to obtain fusion energy: either by means of magnetically confined plasmas, or by inertial confinement. We shall only discuss inertial confinement, as it is only this approach that might employ an accelerator. Inertial confinement may be achieved by using a laser or a particle beam of energetic ions impinging on a pellet of tritium and deuterium from several sides.

The laser method has been rigorously studied as part of thermonuclear weapons programs in a number of countries. At present there are two large facilities; one in France and the other, the National Ignition Facility (NIF) at Livermore, in the US. Both are expected to have a net yield of fusion energy, but the primary purpose of these facilities is not commercial power, but the study of the innards of nuclear weapons. It is as yet unclear whether or not laser implosion will ever be a commercial producer of fusion power.

The second method to create pellet implosion is to drive the pellet with ion beams from an accelerator. The ions must not be too energetic or the mean free path for energy deposition would be larger than the pellet. A good choice is a heavy ion with atomic number of about 100, accelerated to 3 GeV total energy (30 MeV/nucleon). A beam of 4000 A peak current and a pulse duration of 10 nanoseconds is needed in order to achieve the implosion and heating of the pellet. For a practical device, of course, the repetition rate needs to be a number of times per second, and the simplest approach seems to be an induction linac. Typically the pellet is hit from a number of different directions, but there is another alternative, which we know will work, called "indirect drive" where the ion beam is employed to make a bath of x-rays, which then implode the pellet. (This is the way hydrogen bombs work: the primary makes x-rays, which are used to implode a secondary.)

The necessary R&D to achieve a pulse of 3 GeV heavy ions in a 4 kA pulse, 10 ns long, has been the subject of research in Berkeley since the mid 1970s. There have been no "show stoppers." (In the vernacular of the scientific workshop this is something that causes one to think again rather than, as in show business, to applaud.) To obtain such a beam is not, however, easy and a great deal has to be learned about how to handle the very intense space-charge-dominated beams.

A typical configuration starts with a multiple-beam ion source (more than 100 beams) and an injector that accelerates a 20 microsecond pulse of 1 A to 2–3 MeV. There follows acceleration; first with electro-

Energy

In the long-term there are only three possible ways to satisfy the energy needs of mankind: solar energy, breeder reactors, and fusion. Solar energy is a renewable energy and includes bio-mass conversion, solar heating of water, solar cooking, and direct generation of electricity by photovoltaic cells. One should also include wind energy in this list since this is generated by the action of the sun. It is unclear, at this time, if the energy demands of the world can be met with solar energy alone, although clearly solar can, and will, contribute to our energy needs.

Although the cost of photo-voltaic cells has come down, thanks to R&D, it is still a factor of five times more expensive to generate electricity this way than by burning coal.

Breeder reactors confront us with problems of safety, nuclear proliferation and waste storage although, as described in the next section, accelerators can help to alleviate these concerns to some degree. At the present time there are no operational breeder reactors, although there have been efforts in a number of countries to build them. They, too, can be designed so that they alleviate the waste storage difficulty.

Nuclear fusion offers a virtually inexhaustible supply of energy. It "burns" deuterium, which is abundant and readily extracted from sea water. The spent fuel contains no radioactive wastes to be disposed of and there is no chain reaction which might give rise to safety issues and while some components of the reactor will inevitably become radioactive, the half-life of the activity will be much shorter than the products of a normal nuclear reactor. No CO_2 or other air pollution will be produced. At the present time, there are no fusion devices but experimental devices which break even in energy seem very close to realization.

There are two known methods of producing fusion energy. The first is called "magnetic confinement" and uses powerful magnetic fields to confine the burning plasma of deuterium and tritium. (The tritium is generated by the neutrons produced by the fusion, in a lithium "blanket" surrounding the reactor, so only deuterium is needed as a fuel.) Fusion reactors take many forms: stellarators, reversed field pinches, spheromacs, tokamaks, etc. Most of the effort has been on tokamaks and the two largest machines, the Tokamak Fusion Test Reactor (TFTR) in Princeton and the Joint European Torus (JET) in England have produced significant fusion energy. The record, achieved by JET, is 0.8 of the power put in coming out as fusion energy. This should be compared with the ratio of just over 1.0, necessary to break even, and with earlier devices where 10,000 times as much energy was put in as was produced. The next step is to build an international reactor called the International Thermonuclear Experimental Reactor (ITER), first agreed upon by Gorbachev and Reagan, and now, decades later, about to be built. A site for ITER was recently chosen in southern France. ITER is expected to produce more power than it consumes.

The second fusion technique, "inertial confinement," takes a small, very carefully designed pellet of deuterium and tritium, a few millimeters across. Beams of lasers or heavy ions converge on the pellet blowing off the outside layer. The force of reaction to the acceleration of this layer propels the inner core inwards like a rocket, compresses it and heats it. The density required is a thousand times that of a liquid and the temperature needed is a billion degrees! Very soon the pellet explodes, but this takes a short time (less than a microsecond) and due to the pellet's inertia (hence the name) this is long enough for it to ignite into a nuclear fire, burning deuterium and tritium, and yielding up to 100 times the energy needed for compression and heating.

The US program and the German programs on heavy ion inertial fusion have been redirected to short-term goals, although fusion still remains the long-term goal. Certainly their support is far from lavish; perhaps we should say "barely adequate." The progress on magnetic fusion using TFTR, built in the mid-1970s and the last major fusion effort of the United States, was very slow and has now been terminated. ITER has been designed and studied for decades, but not yet financed. The total fusion operating budget, in the United States, is only about 250 M$ (compared to a high energy particle physics budget of more than about three times that amount) and shows how little concern there is for fusion energy development.

In fact, the same should be said about all three alternative energy sources. Besides the modest effort on fusion, there is only a small effort put into solar energy and the various nuclear reactor designs which could lead up, eventually, to breeder reactors. It seems there is very little enthusiasm for alternative energy sources while coal, oil, and natural gas remain a less expensive option. Alternative sources will take many years to develop and one hopes for a more enlightened approach in which they receive more attention. There is a good chance that accelerators will play their part in whichever technique emerges.

static focusing, then with magnetic focusing. (At low velocity the electrostatic focusing is more powerful than magnetic focusing.) At 3 GeV there is longitudinal compression that results in a beam of 200 A/beam and a pulse of 200 ns. Then a final neutralized beam (the space charge is so severe as to require that the positive ion beam be neutralized by stationary electrons) is focused both longitudinally and transversely so as to achieve the desired beam for imploding a pellet. (See Fig. 10.10.)

The Heavy Ion Virtual National Laboratory (Berkeley, Livermore, and Princeton) group has performed a number of scaled down experimental demonstrations of crucial elements in this design, but much remains to be done. An Integrated Beam Experiment (IBX) is proposed that would test all crucial aspects of heavy ion inertial fusion. This accelerator would use a potassium beam taken to 10 MeV. The estimated cost is in the range of 100 M$ but support has not yet materialized.

Germany has also been engaged in R&D for heavy ion fusion, aiming to produce a very intense beam with a peak current of 75,000 A at about 10 GeV. The Germans envisage a sequence of several circular

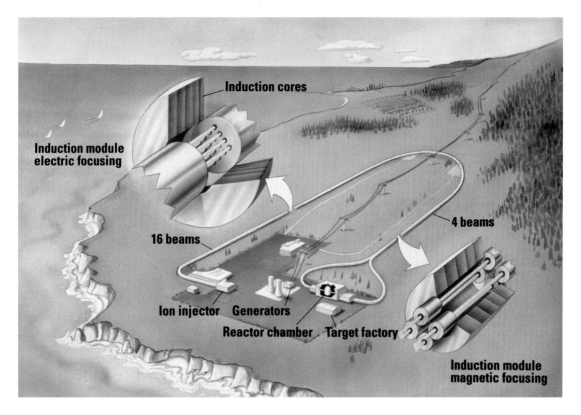

Fig. 10.10 An artist's view of a heavy ion inertial fusion facility. Although the facility is large, it is made of components that appear to be feasible.

machines, in contrast to the Berkeley method of accelerating the ions with a linear accelerator.

Recent years have seen a growing interest in exploring high-energy density plasmas (HEDP). To qualify for this description the plasmas have pressures greater than 1 Mbar, temperatures of more than 1 eV and energy densities of more than 10^{11} J/m^3. (Atmospheric pressure is 1 bar and 1 eV is 10,000 degrees Kelvin.) Such plasmas exist inside planets (where they are warm and dense) and inside stars and other astronomical objects such as the nuclei of active galaxies (quasars), where the plasma is hot and dense. HEDP are, of course, of interest to fusion scientists and nuclear weapon designers, as they need to know the equation of state of various materials under these extreme conditions.

It is interesting to reflect on the astrophysical implications of such research. During many decades scientists have used electrostatic machines and storage rings, where the energy can be very precisely controlled and made much lower than one would have chosen for efficient nuclear reactions, in order to measure the nuclear cross sections relevant to astrophysical conditions. These low energies are typical of the energies with which nuclei approach each other within stars. Thus work on earth has enabled astrophysicists to understand a great

deal about stellar processes and to develop a star model that is very accurate.

Clearly lasers can be used to explore some of the HEDP regions, and the use of the National Ignition Facility (NIF), as well as a number of other laser installations, is under discussion. Ion beams, such as those of RHIC are also of interest. It is of course possible to build an accelerator especially for this, and the group addressing heavy ion fusion has made preliminary studies of the use of an induction accelerator for this purpose.

The other way to obtain fusion energy, magnetic confinement, has been pursued for many years. The latest and most ambitious project, ITER (International Thermonuclear Experimental Reactor), is a multinational effort aiming at "engineering breakeven" (producing more power than the total needed to run the plant) with a large tokamak reactor. A major concern with this approach is the material damage from the 14 MeV neutrons copiously produced in such a reactor. This matter has been discussed for decades but requires a major accelerator for the research and development necessary to resolve it. Such a machine has never been authorized, but the situation must change in the future. There are plans for an International Fusion Materials Irradiation Facility (IFMIF), which would have as its centerpiece an accelerator that would produce copious amounts of 14 MeV neutrons. The cost of the accelerator is estimated to be 2.6 B$.

X.9 Proton Drivers for Power Reactors

At the present time there are no accelerators associated with any aspect of nuclear power reactors; yet this may at some time in the future become one of the most important applications of accelerators. There are at least three different ways in which accelerators might be employed; all require the accelerator to make neutrons, just as in a spallation source. Breeder reactors can achieve some of the same results, but many would find an accelerator (which can be readily shut off) more attractive than a reactor.

The first use of these neutrons might be to breed material suitable for burning in a nuclear reactor. Thorium is plentiful compared to uranium and neutrons convert it into an isotope of uranium that is suitable for fission. India, which has almost no uranium but lots of thorium, has started a research and development program to explore this possibility.

A second use of the neutrons might be to drive a sub-critical reactor: a reactor that could not remain critical without the accelerator beam. There is therefore no danger of a reactor meltdown due to a cooling failure or unplanned release as at Chernobyl: the accelerator is simply shut off. Detailed studies, including experimental

work, have been made in Europe to design such a reactor burning thorium instead of uranium. Thorium, unlike uranium which usually fuels a reactor, cannot produce dangerous, long-lived transuranic isotopes such as plutonium.

A third application is in the "burning up" of long-lived nuclear wastes. This application has received attention in the US, Europe, and Japan. It is possible to simply transmute the transuranic elements (TRUs), and burn long-lived lighter isotopes away, and reduce the necessary isolation time of nuclear wastes from a few hundred thousand years to a few hundred years. Clearly that is a very big difference, and might well make nuclear power attractive in places where it is not currently popular (like the US). After all, the stability of society for hundreds of thousands of years is far beyond anyone's ability to forecast, but a few hundred years seems possible. Such time scales are considered acceptable when storing various chemical wastes.

The idea of combining all three applications has occurred to the Los Alamos group, and also a group in CERN under the direction of Carlo Rubbia (see sidebar in Chapter VI). (See Fig. 10.11.)

We have seen how high power accelerators are needed for a number of very interesting applications, so the motivation for accelerator physicists and engineers is there, and surely the future will see the development of ever-more-powerful accelerators.

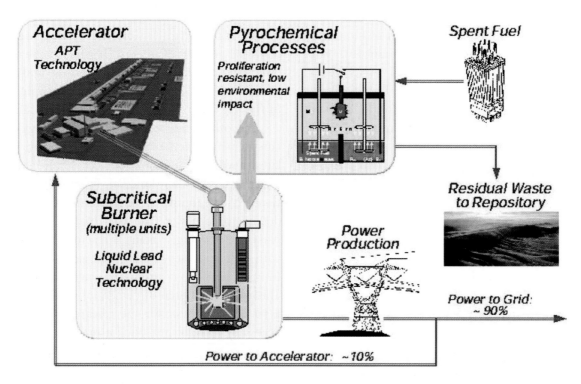

Fig. 10.11 A linac scheme for driving a reactor. These devices can turn thorium into a reactor fuel, power a reactor safely (when the accelerator turns off, so does the reactor) and burn up long-lived fission products so as to reduce the long-term waste storage problem.

X.10 Lasers and Plasmas

If Rutherford were addressing the Royal Society today he might still make relevant and impassioned pleas for a new method to accelerate particles in a reliable and cost-effective way to ten times the present energy. As each new machine has been built, the physics it produces answers some questions but inevitably poses others that only a larger accelerator can answer. The present technology, however, may be close to the end of the road. The plans for linear colliders and the like are so expensive that they will, at best, wait a long time for funds and, at worst, they may well wait forever. What is really needed is not an improvement in present techniques but a *new way* of accelerating. Unfortunately, while bold electrical engineers regularly make advances with relatively well-known techniques, the applied physicist cannot simply dream up a new method by thinking about it — one must await the unpredictable progress of research.

Another difficulty is that big accelerator labs are preoccupied with improving their machines or producing larger versions of them. They are less active in "far out" acceleration development. One must perhaps also look to university departments and to the laboratories that are not immediately involved in the race to high energy. In the US, although every one of the large national laboratories SLAC, Fermilab, Brookhaven, and Lawrence Berkeley still contribute to "future accelerator" research, there is also accelerator research at the Naval Research Laboratory, and at the universities of Cornell, Riverside, UCLA, Texas, Michigan (MSU), Columbia, Yale, and Maryland.

Around the world, there has been activity at Rutherford-Appleton in England, Ecole Polytechnique in France, Osaka and KEK in Japan, the Hebrew University in Israel, the Institute for Laser Engineering in Canada, and the Budker Institute of Nuclear Physics in Russia, and surely, also, at some other places.

It is clear that there is considerable interest in developing new acceleration methods but as yet no "miracle" method has been found — or at least found to be practical. This is sad, for not only would high-energy physics profit from such a "miracle" but it might be much easier to apply the "miracle" to the lower energy devices employed in medical work or in industry.

Accelerating fields in laser beams in vacuum and in laser beams in plasmas may, in fact, be that miracle; the effort in this direction has been both extensive and intensive, during the last two decades. True, no practical device has yet appeared out of this very large R&D effort, but considerable progress has been made and physicists continue to be optimistic that practical accelerators, or significant pieces of overall accelerator systems, will emerge form their efforts.

Ideas for accelerating with lasers emerged in the 50s, soon after the laser was invented and when it was realized that they produce very intense electric fields. There are two major difficulties to be overcome. The first is that a light wave travels at the speed of light and, therefore, is not synchronized with any particles at speeds even slightly less than that of light. The electric field in the light wave will always push the particle one way, and as it overtakes, push it back again so that there is no net acceleration. The second difficulty is that the fields in a light wave are transverse; i.e., the associated electric field (which affects particles) is not in the direction of motion of the light and therefore only good for deflecting particles; not accelerating them. These negative aspects are combined in the Woodyard-Lawson theorem that says that a particle, moving in a straight line in free space, can receive no net acceleration from an electromagnetic wave to first order (directly proportional to the laser field). Therefore, by linear superposition, no combination of light waves, i.e., no shape of light beams, would cause acceleration (see sidebar for Lawson).

In order to accelerate particles at least one of the conditions of the theorem must be violated. For example, one can bring a conductor nearby (so it is no longer free space), but the conductor has to be about a wavelength away, not significantly further. That is easy with the usual acceleration by microwaves, whose wavelengths are in the centimeter range (10 cm for 3 GHz as at SLAC), but not at all for laser light with wavelength of one micron. One can build small structures, but the intense laser fields easily damage the material and put a severe restriction on laser intensity.

A second way, to violate the theorem's assumptions is to have the light propagate through a medium where the light's speed is less than the particle speed. When this happens the particle generates radiation — Cherenkov radiation — and loses energy. That can be run backwards; i.e., use the laser light as input where the Cherenkov radiation would be emitted and so accelerate the particle. The difficulty here is that the medium limits the strength of the laser light (or damage is done to the medium). If the medium is plasma (already ionized) then the laser cannot damage the medium and, as we shall see below, the most interesting laser accelerators are those that employ both a laser and plasma.

A third possibility for violating the theorem's assumptions is to have the particle not move in a straight line. For example, a wiggler could be used and

John Lawson (1923–)

British accelerator physicist
Originator of the Lawson Criterion

It is interesting to see how those that have spent their lives among the Big Machines of Physics found their way into the field. In the case of John Lawson he started at an English grammar school specializing in the most prestigious of subjects — Classics. He was nevertheless fascinated by science and built himself calculators out of cardboard and three-dimensional models of mathematical solids. He had studied enough physics to try for an entrance to Cambridge University (UK) but because a qualification in chemistry was one of the requirements, was "only" accepted into the engineering faculty.

It was war time, and after a brief two-year degree course in mechanical engineering he was sent to the Telecommunications Research Establishment (TRE) Malvern, where the war effort on radar was under way. Thinking that it would lead to an outdoor life, climbing towers, he decided to study antenna design. Although disappointed to find he had to spend most of his time in the laboratory he became fascinated with ways of narrowing down the polar diagram of aerials working with Woodrow and Booker. During this time he hit upon an idea now common on the rooftops of the world — the parabolic dish.

He had little or nothing to do with accelerators at that time but remembers a talk given by linac designer Luis Alvarez on a scheme for covering the leading edge of an aircraft's wing with an array of small dipoles. When war came to an end his expertise in periodic arrays of antennae led him to work at Harwell where a linac was being constructed for measuring nuclear cross sections. He followed rather parallel paths to Alvarez, designing periodic microwave structures. He also had an interest in the first proof of principle synchrotron built by Goward and Barnes using the Woolich betatron as well as the two 30 MeV synchrotrons that followed.

Some of his classified work at Harwell at that time had to do with thermonuclear fusion. Many schemes were then afoot to break even (produce more power out that went in) and in order to review these he invented a criterion: the minimum product of density, confinement time and temperature necessary to break even. This Lawson Criterion remains today the holy grail of fusion research.

In the late 50s he became interested in the new alternating gradient proposals for synchrotrons and found that in the extreme form in which this was first proposed (high Q values) particle resonances from magnetic errors would destroy the beam. He tells how in Europe the term Q for betatron "frequency" was chosen in preference to the US "nu" value at a discussion in John Adam's office. "Q" is not really a frequency but dimensionless, while "nu" is a Greek symbol for a frequency — in radians per second.

It was about this time that the UK had to decide between a high intensity weak focusing machine and the less predictable intensity of a smaller version of the alternating gradient proton synchrotron proposed for CERN. After a "shoot out" in Cockcroft Hall — the Harwell auditorium — Cockcroft himself made the decision for the high intensity Nimrod. Lawson claims his pessimism about resonances was not the principal reason.

Lawson then moved "outside the fence" of Harwell to help build Nimrod at the Rutherford laboratory. He continued to specialize in linac design and microwaves but later played a leading international role in promoting and critically examining ideas for future accelerators. He combines the correct mixture of stimulating the imagination of young people while moderating their over-enthusiasm with the voice of experience.

Now long retired but still with a lively interest in accelerators past present and future he lives between Oxford and Harwell.

---

a free electron laser could be run in reverse in a mode where laser light is fed in to accelerate the electron. This is called an inverse FEL, an IFEL. However it can only operate at energies low enough for the particles to be sensibly bent and is not suitable as a high-energy accelerator, but might be quite useful up to a few hundred MeV.

Still another possibility is to employ the "ponderomotive potential" of the laser, the tendency of a particle to move to where the electric field (and its potential energy) is lowest. The force of this acceleration is only proportional to the square of the laser electric field but, if the laser is powerful enough, this is still of interest. In such accelerators, the electric field of the laser makes the particles jiggle so that their moving charges appear as currents, which, in turn, interact with the laser light's magnetic field to accelerate the jiggling particles.

Although there was much thinking (and a good number of published research papers) about laser accelerators, it wasn't until there was a clear understanding of the Lawson-Woodyard theorem, and, most importantly, the use of plasmas, that the field really took off. It was John Dawson (see sidebar for Dawson) with Toshi Tajima who first suggested the use of plasmas. This theoretical work in 1979 stimulated a Conference in the US on the Laser Acceleration of Particles in 1982, which initiated a series of similar conferences that still goes on today and, most importantly, stimulated the very extensive experimental programs mentioned above.

The basic idea of Dawson and Tajima was the excitation of a plasma electrostatic wave by means of the laser and, then, use of the plasma wave to accelerate particles. The plasma wave has two important features. First, the electric field is longitudinal, in the direction the wave is propagating, and hence good for acceleration. Second, the plasma wave is a slow wave (not one moving at the velocity of light), and the speed can be controlled externally, which makes the wave perfect for accelerating particles. The efficiency of this process, the required stability of the plasma, the injection of electrons to be accelerated, the staging of sections, and the propagation of the laser pulse for long distances are all subjects that have been addressed, with considerable success, in the last decades. There are a number of ways that lasers can excite plasma waves and this serves to characterize the various types of accelerators.

In the Laser Wakefield Accelerator (LWFA) a laser pulse is sent through a plasma with the result that it pushes plasma electrons forward and sidewise. A plasma wave of oscillating electrons is set up after the pulse has gone by, when the electrons return. This is like the wake behind a ship, and the disturbance runs along with the laser pulse, which in a plasma moves at slightly less than the speed of light in free space.

The Plasma Wakefield Accelerator (PWFA) is very similar to the Laser Wakefield Accelerator, but in this case the wake is generated by a tight group of high-energy particles. Coulomb repulsion between the high-energy particles acting on the (low-energy) plasma electrons pushes them aside and so generates the wake. The PWFA has been studied at SLAC (see Fig. 10.12). First experiments showed the focusing effect of the plasma (which was very impressive and might be one of the first uses of plasmas in accelerators). In a 30 cm plasma, using a 28.5 GeV beam as a driver, they were able to accelerate electrons by more than 10 GeV. In very recent work, again using the 28.5 GeV beam, they have observed electrons of 50 GeV! The goal, here, is called an "after burner"; i.e., one pulse of a high-energy beam drives a second pulse to (say) twice the beam energy. For this purpose one needs plasma of 10 m length. The concept, and the theoretical work necessary to carry it out, is due to Thomas Katsouleas.

In the Plasma Beat Wave Accelerator (PBWA) two laser pulses of frequency very close to each other, but with a difference frequency just equal to the oscillation frequency of a plasma electrostatic wave, are fed into plasma. Thus the plasma wave is resonantly excited. In the LWFA the plasma wave is excited by a shock (like suddenly hitting a drum with a sharp blow, after which it resonates at its characteristic frequency). This is in contrast to the PBWA, where the same form of wave is excited resonantly (like giving a swing a small push each cycle until it is swinging with large amplitude and at its characteristic frequency). The first demon-

John Myrick Dawson (1930–2001)

American physicist
Recipient of the American Physical Society's Maxwell Prize and Computational Physics Prize
California Scientist of the Year, 1978

Born in Champaign, Illinois, he received a BS degree in 1952 and a PhD degree in 1957 from the University of Maryland. His PhD thesis was on dense media effects. Upon graduation he joined the Plasma Physics Laboratory in Princeton and remained there from 1957–1973, soon rising to become head of the theoretical group. He spent two years (1969–1970) at the Naval Research Laboratory where he started a group doing plasma simulations. In 1973 he moved to the University of California in Los Angeles and remained there for the rest of his career.

Dawson was a leading figure in the plasma physics community. For more than four decades he made important contributions to science spanning the whole of plasma physics. He was known as the father of computer simulation of plasmas. A visionary, he realized as early as the late 1950s the potential impact of simulations as a way to test both theories and large construction projects before they were built. Dawson used his simulations to test out new ideas such as plasma-based acceleration and by the 1990s, he was realizing his broader vision for simulations in such projects as the Numerical Tokamak.

As a pioneer in the field of plasma-based accelerators he proposed letting particles surf on the plasma-wave wakes left behind by a laser or a particle beam as it moved through plasma. Dawson invented, or was co-inventor of the Plasma Beat Wave Accelerator, the Laser Wake Field Accelerator, the Plasma Lens, the Plasma Wiggler, the Photon Accelerator, and the Ion Channel Laser. In Chapter X.10 the reader will find the very significant experimental progress, in recent years, in these directions.

Dawson was mentor to more than one generation of plasma physicists and encouraged countless others with his insight and bounty of new ideas. For Dawson, science was the most noble of professions and he believed strongly in the importance of controlled nuclear-fusion research, as well as in the many other applications of science for humanitarian needs.

Dawson was known for his eternal optimism and his characteristic smile and chuckle. He issued many wise sayings, which reveal the man as well as his scientific philosophy. One was, "It is always easier to solve a problem if you know the answer before you start," while his favorite rejoinder to a perplexing observation was, "I think we can simulate that!"

stration of a PBWA was at UCLA by Chandrashekhar Joshi, Christopher Clayton and co-workers in 1985 (see sidebar for Joshi). Subsequently work on the PBWA has taken place at Chalk River, École Polytechnique, and Osaka. Typical energy gains of a few tens of MeV have been achieved.

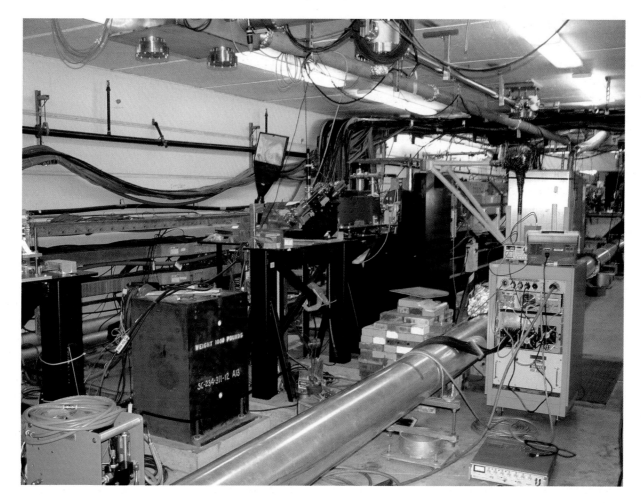

Fig. 10.12 The SLAC End Station first used to study the very fine final focus required for the International Linear Collider. It is in this same area, and using the very intense beam developed for the earlier study, that the experiments on wake-field acceleration were carried out.

Chandrashekhar Joshi (1953–)

Indian-American physicist
APS James Clerk Maxwell Prize for Plasma Physics, 2006
US Particle Accelerator School Prize for Achievement in Accelerator
  Physics and Technology, 1997
American Physical Society Prize for Excellence in Plasma Physics,
  1996
Queen Mary Prize of the Institute of Nuclear Engineers

Born in India, Joshi went to England for both his high school and university education. He received his BS degree from Queen Mary College of the University of London in 1974 and a PhD from Hull University in laser-plasma interactions in 1978. He then spent two years at the National Research Council in Ottawa, Canada. In 1980 he moved to the University of California in

Los Angeles, where he has remained ever since.

Although his experimental work was on traditional plasma problems at UCLA, Joshi was interested in other ideas developed by John Dawson on the laser-plasma acceleration of particles. In 1982, stimulated by the first workshop on the subject, Joshi started an experimental group on laser-plasma acceleration of particles in the Electrical Engineering department at UCLA. This new group had many ties to the theory and especially the numerical simulation work of John Dawson and his co-workers.

Almost all the fine accomplishments described in the section on lasers and plasmas, and those of the more than 30 groups worldwide devoted to this subject, can be traced back to the UCLA Group of Chan Joshi. Joshi himself has become a role model for many. He is currently a distinguished professor of Electrical Engineering at UCLA.

Another idea is the Self-Modulated Laser Wakefield Accelerator (SMLWFA) in which a long intense laser pulse is fed into plasma. Instead of simply exciting a plasma wave (as a short pulse does in the LWFA) the laser pulse slowly generates a plasma wave. Soon the pulse, instead of having a smooth envelope, develops amplitude modulation at the plasma frequency. This modulation is effective for the acceleration of particles. Work on the LWFA and SMLWFA have been done at Livermore, RAL, École Polytechnique, the Naval Research Laboratory, Michigan, Lawrence Berkeley, and KEK. One of the most dramatic results has been the production at LBL of 1 GeV electrons in plasma of only one millimeter in extent. This is an acceleration gradient of 200 GeV/m (which might be compared to that proposed for the ILC of 35 MeV/m; i.e., an increase of a factor of 6,000).

To build an accelerator from an intense laser beam, one has to guide the laser beam for long distances. This is in conflict with the fact that the sharp focus that is required to make a very intense beam extends over only a short distance. The laser light spreads out as one goes away from the focus; the distance over which this spreading occurs is called the Rayleigh length. Beyond this length, some form of focusing or guiding is needed, and much effort has gone into developing plasma channels that, rather like optical fibers, have this property.

Another problem yet to be solved for a plasma/laser accelerator is the injection of electrons in such a way that accelerated electrons have only a very small energy spread. In most of the devices described above all energies from zero to the top energy were present at the same time. However, recent work at the Lawrence Berkeley, École Polytechnique, and RAL, which involves elaborate injection and guiding schemes, has succeeded in producing beams in the range of hundreds of MeV with a narrow energy spread.

There is another, very simple, but most effective way to accelerate particles with a laser and plasma. This is a "one-shot" process that cannot be repeated and, therefore, cannot be used to bring particles to very high energy. A powerful laser is simply pointed at a sheet of solid material; a cloud of electrons comes out the other side. These electrons pull out protons and ions, so the net effect is to produce a pulse of ions that is very short and ranges in energy up to tens of MeV. These ions are already being used for radiological x-ray pictures of imploding pellets and are being considered for use in inertial fusion, generation of medical isotopes, and as injected pulses to an accelerator (conventional or not).

Through the last decades, as described above, there has been world-wide activity on novel acceleration methods. Besides theoretical work and numerical modeling, there has also been experimental work at a good number of locations. However, it seems to us that that laser/plasma accelerators will take at the very least several decades — beyond the two or three that have already gone by — before they can be used in earnest, and they may never match the very high requirements of high-energy physics. It does seem quite possible that these new acceleration techniques will, however, produce machines that will be more convenient, perhaps cheaper to build, and less expensive to operate, for some of the applications described in this book.

Chapter XI. A Final Word

Somewhere in these pages we mentioned that Bob Wilson famously justified accelerators as, "...nothing to do directly with defending our country except to make it worth defending." He probably had in the forefront of his mind the value to civilization of pure scientific discovery — understanding the universe for its own sake. But so cultured a man would certainly have been equally aware of the value of other major contributions made by accelerators to society — their use in other fields of research, their application in industry and in medicine and, not least, their remarkable record in bringing nations together.

XI.1 Understanding the Universe

A deeper understanding of the world around us and the physical laws that govern us cannot fail to help us view our position in that world from a higher and more civilized vantage point. This quest for understanding of our universe is as old as mankind itself. Here are just two of the reasons why pressing into the unknown is important. We are at the point — for the first time in the history of mankind — where we might be able to understand the origin of mass: why the proton and all material bodies are heavy, the electron is lightweight, and the neutrino hardly massive at all. We have discovered that the universe is expanding at an ever-larger rate, and this expansion dramatically shows the conflict between quantum mechanics (a deeply understood theory that is perfectly valid at small distances) and general relativity (also deeply understood and presumably valid at large distances).

XI.2 Applications

In Chapters VIII and IX we described accelerators that have led to nuclear medicine, nuclear scans (PET), radioactive dating, as well as sychrotron light sources for the study of surfaces and study of important biological systems and molecules. These are immediate applications and such is the enthusiasm with which countries are funding these accelerators that we can safely look forward to their rapid growth. This will lead to the development of ever-better pharmaceuticals. We will gain a better understanding of material surfaces that should allow the design of better catalysts for cleaner air from internal combustion engines. Research using accelerators will help provide efficient fuel cells for hydrogen cars, as well as the development of surfaces for the economic capture of solar energy. Proton accelerators may also play a future role in nuclear waste burn-up and in the production of energy from fusion, while ion accelerators will allow us to learn even more about the atomic nucleus.

But the applications of such knowledge are by their very nature impossible to define. The fruits of scientific discovery — especially very deep discoveries — ripen slowly — often too slowly for those who are impatient for an immediate contribution to our material well-being and prosperity. In the last century we applied the knowledge of the forces which govern electromagnetism, which were discovered in the nineteenth century, to electrify homes, make power generators, electrical grids, radio, TV, and radar. The quantum physics of the early twentieth century changed our lives in the last half of that century with laser surgery and grocery store readouts, transistors, integrated circuits, computers, magnetic resonance imaging, and CAT scans. We cannot know the applications that will come in future decades and future centuries, but surely there will be many.

XI.3 Bringing Nations Together

However, one benefit of our field of research — particle physics and its engines of discovery — is immediate. It

Fang Shouxian (1932–)

Chinese physicist
Director of the Institute of High Energy Physics, 1988–1992
Member of the Chinese Academy of Sciences (CAS), 1996–1998
Chairman, Department of Physics and Mathematics at CAS,
  2004–2008

Fang Shouxian, or Fang, as he is known to his colleagues, is an excellent scientific ambassador for his country. His work on accelerators has given him the freedom to travel abroad and to make friends with foreign scientists even when this was not as easy as it is today. His career in accelerators started in the late 1950s, when he was posted to Moscow to study particle dynamics at the Lebedev Institute and at the Joint Institute of Nuclear Research at Dubna. After a couple of years he returned to China and worked on isochronous cyclotrons and isotope separators.

During China's Cultural Revolution he became isolated from the international world of accelerators. Like many other academics, he was sent away to the countryside for one year to appreciate another way of life. Though this did little for his physics, he learned to cook for several hundred fellow workers — a skill which was later very much appreciated by his western friends when there were laboratory feasts and parties.

As the Cultural Revolution began to fade, Fang was able to travel to the US as part of a scientific delegation. One of us (EW) well remembers his first meeting with Fang at Fermilab in 1973. At that time China was a closed community and what little the West had learned from the media about China was larded with negative propaganda. I remember how exciting it was to discover that China's scientists were as open and friendly as one's own colleagues and it was this that encouraged us to collaborate later, when the time came.

The opportunity arrived in 1978, when China opened its doors. There was a plan to build a large proton synchrotron close to the Ming Tombs. Fang was in the forefront of this initiative, visiting CERN to discuss their plans with John Adams and to invite advisors to visit Beijing to train a new generation of Chinese accelerator builders. He and his colleagues at the Institute for High Energy Physics (IHEP) also invited advisors from laboratories in the US and very soon many Western scientists had been educated in Chinese culture and acquired a first-hand experience of the enthusiasm and professional intelligence of their Chinese counterparts.

The next step in deepening the collaboration was for China to send visitors to work at Western labs. In 1982 Fang came to CERN to join the Antiproton Accumulator Group where he designed a new collector ring, inventing a new way of correcting the chromatic behavior of its optics. He and other Chinese physicists who followed him, Guo Wen-Zie, Chen Si-Yu and Chen Bo-Fei, were outstanding in their mastery of mathematical techniques. At that time, computer simulation had already begun to replace the more revealing classical analytic methods in the West. They seemed to us to be custodians of a lost art, determined to commit all hours under the sun to the cause of science. At that time they had to leave their families behind in China and this encouraged the families of Western physicist to welcome them into their own homes — further strengthening lasting bonds of international friendship.

Once the collaboration between China and CERN was in full bloom, it was time for Fang to go to Beijing where he led the design of the Beijing Electron Positron Collider (BEPC) and when this was completed, on schedule and on time, he became its Director.

He had often confided in colleagues that he would have been happy to leave administrative responsibilities to return to theory — but this was not to be — he was too valuable to Chinese Science. In 1988 he became Director of the Institute of High Energy Physics in Beijing. In 1991 he was elected Member of the Chinese Academy of Sciences where he was to become Chairman of Mathematics and Physics. But in spite of these honors he remains that modest and cheerful friend who is always happy to find time to be hospitable to colleagues and talk about experiences shared in years past.

---

provides an effortless opportunity for nations to forget their differences. The scale of the larger of these engines of discovery is such that nations have no choice but to combine their resources to produce them. They must agree on ways to organize and fund such international activity. The endeavor of building these machines then brings together young people from many nations. It teaches them that — in a world where reason is the ultimate test of validity for any new idea, and where politics must be therefore be subservient — there need be no barrier between nationality, race or creed.

In the formation of this community of scientists, the prejudices which separated the warring states of Europe were the first to crumble. Those who came to CERN in the 1950s immediately shed the propaganda which had clothed their thinking for a decade. Even during the darker days of the cold war, scientific contacts with Russia were maintained in the field of accelerators and particle physics. As Europe has unified and the iron curtain has fallen, new states have looked to particle physics research for their first tentative step towards a broader political union. As the US has sought closer ties with the nations of the Pacific Rim, accelerators have been an obvious conduit for new activity.

Among the first visitors to China, when it emerged from the yoke of the Cultural Revolution in 1978, were accelerator builders and physicists forging links which brought many young Chinese to Europe, Japan and the USA. This process of breaking down the boundaries that divide nations is still pursued in the efforts to establish a laboratory in the Middle East — SESAME.

In the earlier pages of this book we described the international atmosphere at a major scientific laboratory — the Cavendish in Cambridge. That spirit — amplified a hundred fold — is to be found in the major high-energy physics laboratories of the world — CERN, Fermilab and KEK. Anyone visiting any of these large modern engines of discovery will find a utopian mixture of young people from the whole world. They number in their thousands and, by joining in the common endeavor of science, have long forgotten that which might separate them. As we look forward to the next engine of discovery, we wonder in which continent the next major high-energy physics facility will be established. We are confident that wherever this is, the young research scientists who will flock there will enrich a more rational world.

XI.4 A Word Especially for the Young

In this book we have sketched the history of particle accelerators from their very beginning, back in the 1920s, through the present and on into the future. Perhaps, this was a passing moment in the history of mankind, but it is much more likely to be an activity that will continue for many decades to come, producing devices not only for physics, but for an ever increasing catalogue of other applications enriching our everyday lives.

We hope some of our readers, who may be already contributing to this adventure, will have enjoyed the book as background to their research field while others, not so involved, will be motivated to participate. As two people who have been involved most of our lifetimes, we speak from experience when we highly recommend this exciting occupation to the lively minds of the present.

Appendix A. Bibliography and References

Books and Reviews

Breizman, B.N. and Van Dam, J.W., editors (1994), *G.I. Budker, Reflections & Remembrances*, American Institute of Physics Press, New York

Bromley, D.A. (1974), "The Development of Electrostatic Accelerators," *Nuclear Instruments and Methods* **122**, 1–34

Bruck, H. (1966), *Accélérateurs circulaires de particules*, Presses Universitaires de France, Paris

Close, F., Marten. M. and Sutton, C. (2002), *The Particle Odyssey: A Journey to the Heart of Matter*, Oxford University Press, Oxford

Freund, H.P. and Parker, R.K. (1989). "Free Electron Lasers," *Scientific American*, April 1989, p. 56

Galison, P. (1997), *Image and Logic: A Material Culture of Microphysics*, University of Chicago Press, Chicago

Heilbron, J.L. and Seidel, R.W. (1989), *Lawrence and His Laboratory*, University of California Press, Berkeley

Herken, G. (2002), *Brotherhood of the Bomb*, Henry Holt and Co., New York

Joshi, Chandrashekhar (2006), "Plasma Accelerators," *Scientific American*, February 2006, p. 41

Kerst, D.W. (1946), "Historical Development of the Betatron," *Nature* **157**, 90–95

Paris, Elizabeth (2004), "Do You Want to Build Such a Machine?: Designing a High Energy Proton Accelerator for Argonne National Laboratory," Argonne National Laboratory Report ANL/HIST-2, Chicago

Pellegrini, C. and Sessler, A.M. (1995), *The Development of Colliders*, American Institute of Physics Press, New York

Rhodes, R. (1986), *The Making of the Atomic Bomb*, Simon and Schuster, New York

Supple, C. (1999), *Physics in the Twentieth Century*, Harry Abrams, New York

Wideroe, R. (1994), *The Infancy of Particle Accelerators*, DESY 94-039

More Technical Accelerator Books

Bryant, P.J. and Johnsen, K. (1993), *The Principles of Circular Accelerators and Storage Rings*, Cambridge University Press

Chao, A. (1993), *Physics of Collective Beam Instabilities in High Energy Accelerators*, John Wiley, New York

Conte, M. and MacKay, W.W. (1991), *An Introduction to the Physics of Particle Accelerators*, World Scientific

Dikanski, N.S. and Pestrikov (1994), *The Physics of Intense Beams ands Storage Rings*, American Institute of Physics, New York

Edwards, D. and Syphers, M.J. (1993), *An Introduction to the Physics of High Energy Accelerators*, Wiley-Interscience

Humphries, S., Jr. (1986), *Principles of Charged Particle Acceleration*, John Wiley, New York

Humphries, S., Jr. (1990), *Charged Particle Beams*, John Wiley, New York

Kolomenski, A.A. and Lebedev, A.N. (1962), *Theory of Circular Accelerators*, John Wiley, New York

Lawson, J.D. (1988), *The Physics of Charged Particle Beams*, Clarendon Press, Oxford

Lee, S.Y. (1999), *Accelerator Physics*, World Scientific, Singapore

Livingood, J.J. (1961), *Principles of Circular Accelerators*, van Norstrand

Livingston, M.S. and Blewett, J.P. (1962), *Particle Accelerators*, McGraw-Hill, New York

Reiser, M. (1994), *Theory and Design of Charged Particle Beams*, John Wiley, New York

Rosenzweig, J. (2003), *Fundamentals of Beam Physics*, Oxford University Press

Scharf, W. (1986), *Particle Accelerators and Their Uses*, *Parts I and II*, Harward Associates, New York

Scharf, W.(1989), *Particle Accelerators — Applications in Technology and Research*, Research Series Press

Wangler, T. (1998), *Principles of RF Accelerators*, John Wiley, New York

Wiedemann, H. (2003), *Particle Accelerator Physics I and II: Basic Principles and Linear Beam Dynamics*, Springer-Verlag, New York

Wille, K. (2000), *The Physics of Particle Accelerators*: *An Introduction*, Oxford University Press

Wilson, E.J.N. (2001), *An Introduction to Particle Accelerators*, Oxford University Press, Oxford

Wilson, R.R. and Littauer, R. (1960), *Accelerators*, *Machines of Nuclear Physics*, Anchor Books, Doubleday and Co., New York

References from Scientific Publications (in order of publication dates)

Slepian, J, "X-Ray Tube," US-Pat. 1,645,304, submitted on April 1, 1922, published on Oct. 11, 1927, submitted in Germany too, published 1928

Ising, G., "Prinzip einer Methode zur Herstellung von Kanalstrahlen hoher Voltzahl" (in German), *Arkiv för matematik o. fysik*, **18**, No. 30, 1–4 (1924).

Breit, G. and Tuve, M.A., *Carnegie Institution Year Book* **27**, 209 (1927/28) (on betatron tests)

Breit, G., Dahl, O., Hafstad, L.R. and Tuve, M.A., (Carnegie Institution), "On Tesla-coil high voltage devices," *Nature* **121**, 535 (1928); *Phys. Rev.* **35**, 51 (1930); **35**, 66 (1930); **35**, 1406 (1930); **36**, 1261 (1930)

Wideröe, R., "Uber ein neues Prinzip zur Herstellung hoher Spannungen," *Arch. f. Electrot.* **21**, 387–406 (1928)

Walton, E.T.S., *Proc. Cambr. Phil. Soc.* **25**, 469–481 (1929)

Lawrence, E.O. and Edlefsen, N.E., *Science* **72**, 376 (1930)

Graaff, R.J. Van de, *Phys. Rev.* **38**, 1919A (1931). First of a series of papers on electrostatic high voltage generators

Lawrence, E.O. and Sloan, D., *Proc. Nat. Acad. Sci.* **17**, 64 (1931) and "The Production of Heavy High Speed Ions without the Use of High Voltages," *Phys. Rev.* **38**, 2022 (1931); reprinted in [Li66], p. 151

Lawrence, E.O. and Livingstone M.S., (first cyclotron in operation; 80 keV, 13 cm), *Commun. Am. Phys. Soc.*, in May 1931; *Phys. Rev.* **37**, 1707 (1931)

Livingston, M.S., "The Production of High-velocity Hydrogen Ions without the Use of High Voltages," Ph.D. thesis, Univ. of California (1931)

Van der Graaf, R.J. (1931), "A 1,500,000 Volt Electrostatic Generator," *Phys. Rev.* **38**, 1919–1920

Cockcroft, J.D. and Walton, E.T.S., "Experiments with High Velocity Ions," **A136**, 619 (1932); **A137**, 229 (1932); **A144**, 704 (1934) *Proc. Roy. Soc. (London).*

Ising, G., "Högspänningsmetoder för atomsprängning" in *Kosmos*, Annuary of the Swedish Phys. Assoc. (1933)

Kerst, D.W., "Acceleration of Electrons by Magnetic Induction," Letter to the Editor of Oct. 15, 1940, *Phys. Rev.* **58**, 841 (1940) and "Induction Electron Accelerator," *Comm. APS*, Nov. 22–23, 1940, *Phys. Rev.* **59**, 110 (1941)

Kerst, D.W. (General Electric), "Magnetic Induction Accelerator," US-patent No. 2,297,305, submitted on Nov. 13, 1940, issued on Sept. 29, 1942

Kerst, D.W., "The Acceleration of Electrons by Magnetic Induction," *Phys. Rev.* **60**, 47–53 (1941); reprinted in [Li66]

Kerst, D.W. and Serber, R., "Electronic Orbits in the Induction Accelerator," *Phys. Rev.* **60**, 53 (1941); reprinted in [Li66]

Kerst, D.W., "20 MeV Betatron or Induction Accelerator," *J. Sci. Instr.* **13**, 387–394 (1942)

Wideröe, R., "Der Strahlentransformator," *Arch. f. Elektrot.* **37**, 542–555 (1943), submitted on Sept. 15, 1942

Veksler, V.I., "A New Method of Accelerating Relativistic Particles," *Comptes Rendus (Dokaldy) de lAcademie Sciences de lURSS*, **43**(8), 329–331 (1944)

McMillan, E.M., "The Synchrotron — A Proposed High Energy Particle Accelerator," *Phys. Rev.* **68**, 143–144 (1945) (Issue Nos. 5 and 6, Sept. 1 and 15, 1945), submitted on Sept. 5, 1945

Veksler, V., "A New Method of Acceleration of Relativistic Particles," *Journal of Physics UdSSR* **9**, 153–158 (1945), English translation in [Li66]

Goward, F.K. and Barnes, D.E., *Nature* **158**, 413 (1946)

Alvarez, L.W., "The Design of a Proton Linear Accelerator," *Phys. Rev.* **70**, 799 (1946)

Goward, F.K. and Barnes, D.E., *Nature*, **158**, 413 (1946)

Wilson, R.R.. "Radiological Use of Fast Protons," *Radiology* **47**, 487 (1946)

Elder, F.R., Gurewitsch, A.M., Langmuir, R.V., and Pollock H.C., "A 70 MeV Synchrotron," *J. Appl. Phys.* **18**(9) 810–818 (1947).

Christofilos, N.C., unpublished report and U.S. Patent no. 2.736,799, filed March 10, 1950, issued Feb. 28th, 1956

Courant, E.D., Livingston, M.S., and Snyder, H.S., "The Strong-Focusing Synchrotron — A New High-energy Accelerator," *Phys. Rev.* **88**, 1190–6 (1952)

Wideröe, R., "Das Betatron," *Z. f. angewandte Physik* **5**, 187200 (1953) (with many references)

Kerst, D.W., Cole, F.T., Crane, H.R., Jones, L.W., Laslett, L.J., Ohkawa, T., Sessler, A.M., Symon, K.R., Terwilliger, K.W. and Nilsen, N.V., "Attainment of Very High Energy by Means of Intersecting Beams of Particles" (submitted Jan. 23, 1956), *Phys. Rev. Lett.* **102**, 590–591 (1956)

O'Neill, G., "The Storage Ring Synchrotron," *Proc. Int. Conf. on High Energy Accelerators*, Vol. 1, 1956, pp. 64–67

Symon, K.R. et al., "Fixed Field Alternating-Gradient Particle Accelerators," *Phys. Rev.* 103, 1837 (1956)

Courant, E.D. and Snyder, H.S., "Theory of the Alternating-Gradient Synchrotron," *Annals of Physics*, No. 3, 1–48 (1958)

O'Neill, G., "Storage Rings for Electrons and Protons," *Proc. Int. Conf. on High-Energy Accelerators and Instrumentation*, CERN, Geneva, 1959, pp. 125–136

Touschek, B., First proposal to built an electron-positron storage ring, presented at Frascati, March 7, 1960 [Am81]

Budker, G.I., "Creation of Accelerators with Colliding Beams," *Vestnik Academii Nauk SSSR*, No. 6, 1964

Johnsen, K. (ed.) *The Design Study of Intersecting Storage Rings (ISR)*, CERN AR/Int. SG/64-9 (1964)

Livingston, M.S., "The Development of High-Energy-Accelerators," commented reprints or translations of original papers (book), Dover, New York, 1966

Scharf, W.H. and Chomiki, C.A., *Physica Medica* Vol. XII No. 4, 199 (1966)

Budker, G.I., "An Effective Method of Damping Particle Oscillations in Proton and Antiproton Storage Rings," *Sov. J. Atom. Energy* **22**, 438 (1967)

Wilson, R.R., "The NAL Proton Synchrotron," *Proc. 8th Int. Conf. High Energy Accelerators*, CERN, 1971

van der Meer, S., "Stochastic Damping of Betatron Oscillations in the ISR," CERN/ISR-PO/72-31, 1972

Wilson, E.J.N. (ed.), "The 300 GeV Programme," CERN/1050, 1972

Blewett, M.H., "Theoretical Aspects of the Behaviour of Beams in Accelerators and Storage Rings," CERN 77-13, 1977, pp. 111–138

Degèle, D., PETRA, *Proc. 8th Int. Conf. High Energy Accelerators*, Geneva, 1980

Richter, B. et al., *SLC Design Report*, SLAC 2590, 1980

Amaldi, E., "The Bruno Touschek Legacy," CERN-Report No. 8119, 83 pages, 1981

HERA Design Team, 1981 "HERA — A Proposal for a Large Electron Proton Colliding Beam Facility at DESY," DESY HERA 81/10, 1981

Tajima, T. and Dawson, J.M., Proc. Workshop on Laser Acceleration of Particles, Los Alamos (AIP Conf. Proc. No. 91), 1982, p. 69

Loew, G.A., "Elementary Principles of Linear Accelerators," SLAC PUB 3221. Also in AIP Conf. Proc. No. 105, 1983

Billinge, R., "Design and Operation of the CERN SppS Collider: Overview," *Proc. 1984 Summer School on High Energy Particle Accelerators*, FNAL, 1984

CERN, Geneva, "LEP Design Report," CERN-LEP/84-01, 1984

Evans, L., "Design and Operation of the CERN SppS Collider," *Proc. 1984 Summer School on High Energy Particle Accelerators*, FNAL, 1984

Dawson, J.M. and Chen, P., SLAC-PUB-3601, 1985

van der Meer, S., CERN-PS-85-65, 1985

Lefèvre, P. (ed.), *The Large Hadron Collider*, CERN/AC/95-05, 1995

Prior, C.R., "Status of the HIDIF Study," *Proc. 6th European Particle Accelerator Conference – EPAC 98*, Stockholm, Sweden, 1998

Delahaye J.-P. et al., CERN/PS 99-005 (LP), PAC, New York, 1999

O'Shea, P. and Freund, H. "Free Electron Lasers: Status and Applications," *Science* **292**, 1853 (2001)

Web Addresses

The Spallation Neutron Source, http://www.sns.gov/aboutsns/aboutsns.htm

The International Linear Collider, http://www.linearcollider.org/cms/

The Linear Coherent Light Source, http://www-ssrl.slac.stanford.edu/lcls/

National Association for Proton Therapy, http://www.proton-therapy.org/

The Large Hadron Collider, http://.web.cern.ch//

The Compact Linear Collider, http://clic-study.web.cern.ch/CLIC-Study/

CERN, http://public.web.cern.ch/public/

Lawrence Berkeley National Laboratory, http://www.lbl.gov/

SLAC, http://www.slac.stanford.edu/

Fermilab, http://www.fnal.gov/

Brookhaven National Laboratory, http://www.bnl.gov/world/

Argonne National Laboratory, http://www.anl.gov/

Oak Ridge National Laboratory, http://www.ornl.gov/

KEK, http://www.kek.jp/intra-e/

DESY, http://www.desy.de/html/home/index.html

GSI, http://www.gsi.de/

Budker Institute of Nuclear Physics, http://sky.inp.nsk.su/index.en.shtml

European Synchrotron Radiation Facility, www.esrf.fr

Institute of High Energy Physics, www.ihep.ac.cn/english/index.html

Japan Synchrotron Radiation Research Institute (JASRI), www.spring8.or.jp/en

Jefferson Laboratory, www.jlab.org

Los Alamos National Laboratory, www.lanl.gov

Paul Scherrer Institute, www.psi.ch/

Rutherford Appleton Laboratory, http://www.cclrc.ac.uk/Activity/RAL

Appendix B. The Accelerator Community

It is impossible for us, much as we would like to, to list all who have contributed, and are, contributing to accelerator physics. The Particle Accelerator Conference, in the US, attracts an attendance of 1,200, while the European Particle Accelerator Conference about 1,000 participants and the Asian Accelerator Conference about 500. Typically only a fraction of accelerator physicists are able to attend a Conference; perhaps one-third of active scientists; and it is therefore reasonable to assume that the number of accelerator physicists and engineers must be over 7,500. If we include those who are no longer active the total community probably numbers more than 10,000 accelerator scientists.

There are also professional societies: the American Physical Society, the European Physical Society, etc. Some of these Societies hold special accelerator meetings; the Russian Academy of Sciences has been doing that for many years (and typically 500 attend each meeting).

The American Physical Society bestows the R.R. Wilson Prize for outstanding contributions to accelerator science. The recipients, through the years (in inverse chronological order with the most recent first) were: Glen Lambertson, Keith R. Symon, Katsunobu Oide, John Seeman, Helen T. Edwards, A.N. Skrinsky, Claudio Pellegrini, Maury Tigner, Robert B. Palmer, Mathew Sands, Andrew M. Sessler, Albert Josef Hofmann, Raphael M. Littauer, Gustav-Adolf Voss, Thomas L. Collins, John P. Blewett, Rolf Wideröe, J. Reginald Richardson, Kjell Johnsen, Alvin V. Tollestrup, Martin N. Wilson, Donald W. Kerst, and Ernest D. Courant.

The US Particle Accelerator School presents a Prize for Achievement in Accelerator Physics and Technology.

Previous recipients (in inverse chronological order) were: Anton Piwinski, Wim Leemans, Martin Reiser, Sami Tantawi, Tor Raubenheimer, Dieter Möhl, Bruce Carleston, Robert Gluckstern, Daniel Boussard, Chandrashekhar Joshi, Herman Winick, James Spencer, Tsumoru Shintake, Richard Sheffield, John Fraser, Marc Ross, Glen Lambertson, Wolfgang Schnell, Rolf Wideroe, Donald Prosnitz, Matthew Sands, Daniel Birx, Karl Brown, I.M. Kapachinskii, V.A. Teplyakov, Andrew Sessler, Klaus Halbach, Lars Thorndahl, Helmut Piel, Maury Tigner, Thomas Weiland, Helen Edwards, John Madey, Ernest Courant, M. Stanley Livingston, and Robert Wilson.

The European Physics Society presents an Accelerator Prize, and previous recipients (again in inverse chronological order; most recent last) are: Hakan Danared, Igor Syrachev, Jeffrey Hangst, R.D. Kohaupt, The DESY Feedback Group, Soren Pape Moller, Crtoforo Benvebuti, Pantaleo Raimondi, Eberhard Keil, Frank Zimmermann, Kurt Hubner, Vladimir Shiltsev, Igor Meshkov, and Vladimir Teplyakov.

The IEEE awards a Particle Accelerator and Technology Award, and previous recipients have been (in chronological order): L. Jackson Laslett, Perry Wilson, Z.D. Farkas, Ronald Scanlon, David Larbelestier, Thomas Collins, Louis Anderson, Yoshiharu Mori, Pierre Lapostolle, Jtirgen Struckmeier, K. Leung, David Sutter, Ilan Ben-Zvi, G. William Foster, Gerald Jackson, John Sereman, Lloyd Young, Keith Symon, Stephen Milton, Ronald Davidson and Thomas Roser.

Appendix C. Glossary

ADA	Ch VI	A colliding beam machine
Adone	Ch VI	A colliding beam machine
AGS	Ch V	A synchrotron
Alternating Gradient	Ch V	A method of focusing. Also called strong focusing
ampere (A)		A unit of current. A 100 watt electric bulb on a 110 volt line has almost 1 A passing through it
anti-proton (p-bar)	Ch VI	A particle with charge -1 and a mass 1836 times that of an electron
Argonne National Laboratory (ANL)	Ch VIII	A laboratory near Chicago, Illinois
atto-second		10^{-18} s
BESSY	Ch VIII	A synchrotron radiation machine
betatron	Ch IV	A type of circular accelerator
Bevalac	Ch III	The Super Hilac combined with the Bevatron
Bevatron	Ch V	A circular accelerator at Lawrence Berkeley National Lab.
brachytherapy	Ch IX	Insertion of radioactive pellets into a tumor
bremsstrahlung	Ch III	Radiation of x-rays when electrons slow down in material
Brookhaven National Laboratory (BNL)	Ch V	A laboratory on Long Island, New York
calutron	Ch II	A device for separating isotopes of uranium

Cambridge Electron Accelerator (CEA)	Ch VIII	A laboratory in Massachusetts
CEBAF	Ch III	The superconducting accelerator located at Jefferson laboratory
Cerenkov counter	Ch VII	Radiation emitted by a particle in a medium
CERN	Ch V	Laboratory in Switzerland
CESR	Sidebar Ch VI	A colliding beam machine at Cornell
chips	Ch I	Computer elements
Cockcroft-Walton	Ch I	An type of electrostatic accelerator
coherent instability	Sidebar Ch VI	Particles in a bunch acting together
coherent radiation	Ch VIII	Electromagnetic radiation where all the photons move in lock-step (like laser light vs. electric bulb light)
collider	Ch VI	A device for colliding beams head-on
cooling of beams	Ch X	Reduction in the transverse velocity spread and/or reduction in the energy spread of a beam (without changing the size of the beam)
Cosmotron	Ch V	A synchrotron
CP	Ch VI	Charge-Parity; a symmetry
CPT	Ch VI	Charge-Parity-Time Reversal; a symmetry
cyclotron	Ch II	A circular accelerator
Dees	Ch II	The metal shaped structures that provide the acceleration in a cyclotron
DESY	Ch VI	A laboratory in Germany
Dewar	Ch V	A cryogenic vessel that provides excellent insulation
dipole	Ch V	A magnetic element with two-fold symmetry
Doppler effect		The increase of frequency when moving towards a source and a decrease in frequency when the source recedes
DORIS	Ch VI	A colliding beam machine

Drift Tube Linac (DTL)	Ch III	A type of linac most suitable for accelerating slow moving particles
electrostatic	Ch I	A type of accelerator
electron cloud	Sidebar Ch VI	A coherent instability
electron cooling	Ch I	A method of cooling an ion beam by making it interact with a cold electron beam
Electron Ring Accelerator	Ch III	A collective accelerator; a Smokatron
European Union X-Ray Free Electron Laser (EU-XFEL or FLASH)	Ch VIII	The synchrotron source at DESY (2006)
eV		A unit of energy equal to that of an electron accelerated by a one volt potential
FEL	Ch VIII	A free electron laser
femto-second		10^{-15} s
Fermilab	Ch V	A laboratory in Illinois
FFAG	Ch II	An accelerator with fixed (in time) alternating (in space) magnetic field
focusing	Ch II	Transverse or longitudinal forces that hold the particles being accelerated from deviating from their proper orbit
gamma rays		Electromagnetic radiation of energy (approximately) greater than 1 MeV
Gesellschaft für Schwerionenforschung (GSI)	Ch X	A laboratory in Germany
GeV		A unit of energy equal to that of an electron accelerated by a 10^9 volt potential
Gigaherz		A thousand million times a second
glioblastoma	Ch IX	A tumor of the brain
hadron	Ch VI	An elementary particle of heavy mass such as a nucleon

head-tail	Sidebar Ch VI	A type of coherent instability
HERA	Ch VI	A collider of electrons and protons at DESY
Higgs	Sidebar Ch VI	An intermediate particle believed to be the source of mass. Not yet observed
HILAC	Ch III	An ion linear accelerator at LBNL
HIMAC	Ch IX	A cancer therapy facility in Japan
induction unit	Ch III	An acceleration unit in which the particle beam is like the secondary in a transformer
Intermediate Vector Bosons	Sidebar Ch VI	Particles that carry the weak force
ionization cooling	Ch X	A beam cooling technique which relies on the beams ability to ionize material it passes through
ISR	Ch VI	A proton-proton colliding beam machine at CERN
Japanese Linear Collider (JLC)	Ch X	The Japanese concept for a linear collider
Jefferson Laboratory	Ch III	A laboratory located in Virginia
Joules		A unit of energy
KEK	Ch VI	The high-energy laboratory of Japan
keV		A unit of energy equal to that of an electron accelerated by a thousand volt potential
kHz		A thousand times a second
kilo-volts (kV)		One thousand volts. Many of the larger power transmission lines run at 115 kV
Landau damping	Sidebar Ch VI	A mechanism for damping coherent instabilities
laser cooling	Sidebar Ch VI	Cooling by using laser Doppler effect
Lawrence Berkeley National Laboratory (LBL or LBNL)	Ch I	A laboratory in California
L-Band	Sidebar Ch VI	A radar frequency band containing the very popular linac frequency of 1.3 GHz

LBWA	Ch X	Laser Beat Wave Accelerator
LEP	Ch VI	A circular collider of electrons and positrons at CERN
lepton		A light particle such as an electron
	Ch VI	A circular collider of protons at CERN
linac	Ch III	A linear accelerator
Linac Coherent Light Source (LCLS)	Ch VIII	The Stanford free electron x-ray laser
LWFA	Ch X	Laser Wake Field Accelerator
Magnetic Resonance Imaging (MRI)	Ch V	Nuclear spin resonance used to non-invasively image biological material
Material Testing Accelerator (MTA)	Ch III	The very large accelerator built by Lawrence to produce nuclear material
megajoules (Mjoules)		A million joules
megawatt		A unit of power; a million watts
meson	Ch VI	A particle with mass between that of an electron and a proton
meter (m)		A unit of length (39 inches)
MeV		A unit of energy equal to that of an electron accelerated by a million volt potential
MHz		A million times a second
micron		A unit of length; one millionth of a meter
microsecond		A millionth of a second (10^{-6} s)
milliamp (mA)		A thousandth of an ampere
millicuries		A unit of radiation intensity equal to 2.6×10^7 disintegrations per second
millimeter (mm)		A thousandth of a meter
millisecond (ms)		A thousandth of a second

muon	Sidebar Ch X	A particle 210 times heavier than an electron that decays in 2.2 microseconds
nanometer		A thousand millionth of a meter (10^{-9} m)
nanosecond		10^{-9} s
negative mass	Sidebar Ch VI	A type of coherent instability
neutrino	Ch X	A particle of very tiny mass and no charge. There are electron, muon, and tau neutrinos
nucleon	Ch I	A neutron or a proton
ohmic losses	Ch V	Resistive losses on an electric current
Omnitron	Ch III	An unbuilt circular accelerator of all species of atoms
oncologist	Ch IX	A physician who treats cancer
parton	Sidebar Ch VI	A constituent of a nucleon (really just a quark)
Peletrons	Ch I	An electrostatic accelerator
PEP	Sidebar Ch VI	A colliding beam machine at SLAC
PETRA	Ch VI	A colliding beam machine at DESY
Phasotron	Ch V	A synchrotron located in Dubna, Russia
pico-second		10^{-12} s
pion	Ch III	A meson of mass 280 times that of an electron that decays in 2.6×10^{-8} s
plasma	Ch X	A neutral mixture of positive ions and electrons
ponderomotive potential	Ch X	The force on particles which depends upon the square of the electric field
proton (p)	Ch I	A particle with charge +1 and a mass 1836 times that of an electron
Proton Induced X-ray Emission (PIXIE)	Ch I	A proton going through material emitting x-rays
PS	Ch V	A proton synchrotron at CERN
quadrupole	Ch V	A magnetic element with four-fold symmetry

quark	Sidebar Ch VI	Basic constituents of hadrons and mesons
quark-gluon plasma	Ch III	The situation inside a very hot nucleus when quark and gluons are released from the nucleons within the nucleus
radioisotopes	Ch IX	Radioactive species of nuclei
Rayleigh length	Ch X	The length in which a beam of light spreads so that its transverse size is doubled
recirculating linac	Ch VIII	A method to produce synchrotron radiation
reluctance	Sidebar Ch V	A measure of the difficulty of establishing a magnetic field
resistive wall	Sidebar Ch VI	A coherent instability
RF	Ch III	Radio Frequency
RFQ	Ch III	Radio Frequency Quadrupole
rumbatron	Ch III	A radio-frequency cavity
S-band	Ch III	A radar frequency band containing the very popular linac frequency of 3 GHz
scintillators	Ch VII	Materials that give off visible light when a particle moves through it
second (s)		A unit of time
SESAME	Ch VIII	A synchrotron radiation facility in Jordan
SLC	Ch VI	Stanford Linear Collider
smokatron	Ch III	An accelerator using the collective effect of a ring of electrons
solenoidal	Ch VI	A magnetic field along the particle motion
Southeastern Universities Research Association (SURA)	Ch III	The managing organization for the Jefferson laboratory
spallation	Ch X	A proton hitting a heavy nucleus and typically breaking it up into many nucleons
SPEAR	Ch VI	A colliding beam machine at SLAC

SPS	Ch V	A proton synchrotron at CERN
SSC	Ch VI	A proton collider (unbuilt) in the US
stacking	Ch VI	A method of building up an intense beam
Standard Model	Sidebar Ch VI	The theory covering elementary particles
Stanford Linear Accelerator Center (SLAC)	Ch III	A laboratory in California
stochastic cooling	Sidebar Ch VI	Cooling based upon fluctuations
strong focusing	Ch V	A means of focusing. Also called alternating gradient focusing
SuperHILAC	Ch III	A linear accelerator of all atomic species at LBNL
Swindletron	Ch I	An electrostatic accelerator
synchrotron	Ch V	A circular accelerator
TERA	Ch IX	A cancer therapy facility
Tesla (T)		A unit of magnetic field. A 30 pound loudspeaker has a field strength of 1 T; an MRI runs at about 7 T
TeV		A unit of energy equal to that of an electron accelerated by a 10^{12} volt potential
Tevatron	Ch VI	A collider of protons and anti-protons at Fermilab
thermalize	Ch X	When two items (here an ion beam and an electron beam) come to the same temperature
thyratron	Ch III	A radio tube capable of producing high voltage pulses
tokamak	Ch III	A magnetic confinement fusion device
TPC	Ch VII	Time Projection Chamber
transformers	Ch III	A device that transforms current voltage
transuranic	Ch III	Elements beyond uranium
undulator	Ch VIII	A device for repeatedly deflecting electrons
van de Graaff	Ch I	A type of electrostatic accelerator

VEPP	Ch VI	A colliding beam machine in Russia
volt (V)		A unit of electrical strength. Utility plugs in the US run at 110 volts; those in Europe at 220 volts
W	Sidebar Ch VI	A heavy particle carrying the weak force
Wideroe linac	Ch III	A type of linear accelerator
wiggler	Ch VIII	A device for repeatedly deflecting electrons
x-rays		Electromagnetic radiation of energy in the range (approximately) of 100 eV to 100 keV

Appendix D. List of Illustrations with Acknowledgments

Figures

1.1   Courtesy of University of Cambridge, Cavendish Laboratory
1.2   Courtesy of University of Cambridge, Cavendish Laboratory
1.3   Courtesy of Fermilab
1.4   Courtesy of Bromley, D.A., *Nucl. Instrum. Methods* 122, 1 (1974)
1.5   Courtesy of MIT
1.6   Courtesy of MIT
2.1   Courtesy of Lawrence Berkeley National Laboratory
2.2   Courtesy of Lawrence Berkeley National Laboratory
2.3   Courtesy of Lawrence Berkeley National Laboratory
2.4   Courtesy of Lawrence Berkeley National Laboratory
2.5   Courtesy of Lawrence Berkeley National Laboratory
2.6   Collection of Lawrence Jones
2.7   Collection of Lawrence Jones
2.8   Courtesy of Rutherford Appleton Laboratory
2.9   Courtesy of Lawrence Berkeley National Laboratory
3.1   Courtesy of Pedro Waloschek
3.2   Courtesy of Lawrence Berkeley National Laboratory
3.3   Courtesy of Lawrence Berkeley National Laboratory
3.4   Courtesy of CERN
3.5   Courtesy of Varian Inc.
3.6   Courtesy of SLAC
3.7   Courtesy of CERN
3.8   Courtesy of Lawrence Livermore National Laboratory
3.9   Courtesy of Lawrence Livermore National Laboratory
3.10   Courtesy of Los Alamos National Laboratory
3.11   Courtesy of Varian
4.1   Courtesy of J. R. Staples, LBL
4.2   Courtesy of J. R. Staples, LBL
4.3   Courtesy of Russell NDE Systems Inc
5.1   Courtesy of UKAEA
5.2   Courtesy of US Government
5.3   Courtesy of UKAEA
5.4   Courtesy of CERN
5.5   Courtesy of Lawrence Berkeley National Laboratory
5.6   Courtesy of JINR
5.7   Courtesy of CERN
5.8   Courtesy of Brookhaven National Laboratory

5.9	Courtesy of CERN
5.10	Courtesy of Fermilab
5.11	Courtesy of Fermilab
5.12	Courtesy of CERN
5.13	Courtesy of Fermilab
6.1	Courtesy of Stanford University
6.2	Courtesy of Budker Institute
6.3	Courtesy of Lab. Naz. Frascati INFN © 1961 INFN All Rights Reserved
6.4	Courtesy of Lab. Naz. Frascati INFN © 1961 INFN All Rights Reserved
6.5	Courtesy of CERN
6.6	Courtesy of CERN
6.7	Courtesy of CERN
6.8	Courtesy of CERN
6.9	Courtesy of CERN
6.10	Collection of E. Wilson
6.11	Courtesy of CERN
6.12	Courtesy of Fermilab
6.13	Courtesy of DESY
6.14	Courtesy of CERN
6.15	Courtesy of CERN
6.16	Courtesy of Brookhaven National Laboratory (Star Collaboration)
7.1	Courtesy of Fermilab
7.2	Collection of Don Perkins
7.3	Collection of Don Perkins
7.4	Courtesy of CERN
7.5	Courtesy of CERN
7.6	Courtesy of CERN
7.7	Courtesy of Brookhaven National Laboratory
7.8	Courtesy of CERN
8.1	Courtesy of ESRF
8.2	Courtesy of CERN
8.3	Courtesy of ESRF
8.4	Courtesy of SPring 8
8.5	Peter Ginter photo, Courtesy of Stanford Linear Accelerator Center
8.6	Courtesy of Stanford Linear Accelerator Center (2006)
8.7	Courtesy of DESY
8.8	Courtesy of Lawrence Berkeley National Laboratory
9.1	Courtesy of Varian Inc.
9.2	Courtesy of HIMAC
10.1	Courtesy of Stanford Linear Accelerator Center
10.2	Courtesy of KEK
10.3	Courtesy of CERN
10.4	Courtesy of DESY
10.5	Courtesy of Rutherford Appleton Laboratory
10.6	Courtesy of Spallation Neutron Source
10.7	Courtesy of University of Tokyo Super Kamiokande Laboratory
10.8	Courtesy of RIKEN
10.9	Courtesy of the MICE Collaboration
10.10	Courtesy of Lawrence Berkeley National Laboratory Fusion Energy Research Programme
10.11	Courtesy of Los Alamos National Laboratory
10.12	Courtesy of Stanford Linear Accelerator Center

Portraits

Adams	Courtesy of CERN
Alvarez	Courtesy of Lawrence Berkeley National Laboratory
Blewett	Courtesy of Brookhaven National Laboratory
Billinge	Courtesy of Rosemary Billinge
Budker	Courtesy of Budker Institute for Nuclear Physics
Cerenkov	Courtesy of Elena P. Cherenkova
Charpak	Courtesy of CERN
Christofilos	Courtesy of A.C. Melissinos
Cockcroft	Courtesy of CERN
Collins	Courtesy of FNAL
Courant	Courtesy of Brookhaven National Laboratory
Dahl	Courtesy of CERN
Dawson	Courtesy of UCLA
Edwards	Courtesy of FNAL
Evans	Courtesy of CERN
Fang	Courtesy of Fang Shouxian
Ginzton	Courtesy of Varian Inc.
Glaser	Courtesy of the Lawrence Berkeley National Laboratory
Graaff	Courtesy of MIT
Green	Courtesy of Brookhaven National Laboratory
Grunder	Courtesy of Argonne National Laboratory
Halbach	Courtesy of Lawrence Berkeley National Laboratory
Hansen	Courtesy of Stanford Linear Accelerator Center
Herb	Courtesy of the American National Academy of Sciences
Hereward	Courtesy of CERN
Hine	Courtesy of CERN
Hofmann	Courtesy of CERN
Hubner	Courtesy of CERN
Johnsen	Courtesy of CERN
Joshi	Courtesy of C. Joshi
Kerst	Courtesy of University of Wisconsin
Kurokawa	Courtesy of S. Kurokawa
Lawrence	Courtesy of Lawrence Berkeley National Laboratory
Lawson	Courtesy of UKAEA
Leow	Courtesy of G. Leow
Livingston	Courtesy of Brookhaven National Laboratory
Lofgren	Courtesy of Lawrence Berkeley National Laboratory
Madey	Courtesy of J. Madey
McDaniel	Courtesy of Cornell University
McMillan	Courtesy of Lawrence Berkeley National Laboratory
Miller	Courtesy of Stanford Linear Accelerator Center
Nygren	Courtesy of Lawrence Berkeley National Laboratory
Ohkawa	Courtesy of T. Ohkawa
Oide	Courtesy of Oide
Oliphant	Courtesy of CERN, Pauli Archives
O'Neill	Courtesy of Space Studies Institute
Ozaki	Courtesy of S. Ozaki
Palmer	Courtesy of Brookhaven National Laboratory
Panofsky	Courtesy of CERN

Pelligrini	Courtesy of UCLA
Picasso	Courtesy of CERN
Reardon	Courtesy of P. Reardon
Rees	Courtesy of SLAC
Richter	Courtesy of B. Richter
Robinson	Courtesy of G. Voss
Rubbia	Courtesy of CERN
Rutherford	Courtesy of Emilio Segrè Visual Archives
Sands	Courtesy of SLAC
Schnell	Courtesy of CERN
Schopper	Courtesy of CERN
Seeman	Courtesy of SLAC
Skrinsky	Courtesy of A. Skrinsky
Snyder	Courtesy of Brookhaven National Laboratory
Symon	Courtesy of K. Symon
Telnov	Courtesy of V. Telnov
Thomas	Courtesy of Emilio Segrè Visual Archives
Tigner	Courtesy of Cornell University
Tollestrup	Courtesy of FNAL
Touschek	Courtesy of INFN Frascati
van der Meer	Courtesy of CERN
Veksler	Courtesy of CERN
Voss	Courtesy of G. Voss
Walton	Courtesy of Lawrence Berkeley National Laboratory
Wideroe	Courtesy of Pedro Waloschek
Wiik	Courtesy of Robert Palmer
Wilson, M.	Courtesy of M. Wilson
Wilson, R.	Courtesy of FNAL
Winick	Courtesy of H. Winick

Index

A

acceptance 41, 154, 156
AdA 81, 85, 86
Adams, Sir John 68
ADONE 84, 87
AGS (Alternating Gradient Synchrotron) 64, 65
ALEPH, track reconstruction 114
Allibone, Thomas Edward 3
Alvarez, L.W. 31
Amaldi, Ugo 138
antiproton 94, 100
    source 102
ASCLEPIOS 138
Asian Accelerator School 104
Astron 44, 45, 46
ATLAS detector 116, 118
ATLAS, superconducting heavy ion accelerator 44

B

B-factory 40
Bailey, J. 93
Barish, B. 146
Barnes, D.E. 54, 171
beam cooling 155
beat-wave accelerator 130
Bennett, W.H. 8
BEPC (Beijing Electron Positron Collider) 167
Berkeley University 11
BESSY 126
betatron 50–53
    100 MeV 53
    20-inch 52
    after WW II 53
    Kerst 51
    modern 54
    Wideroe 51
beta beams 154
Bethe, Hans 19
Bevalac 98, 107
Bevatron 58
Billinge, Roy 72

Birge, Robert 14
Blewett, John Paul 43
Blosser, Henry 25
BNCT (Boron Neutron Capture Therapy) 138
Booster synchrotron 72
brachytherapy 134
Bragg peak 135, 137
Breit, Gregory 50, 51
Budker, Gersh Itskovich 83
Budker Institute of Nuclear Physics at Novosibirsk 84

C

cancer therapy, future 138
calutrons 19, 21
Cambridge Electron Accelerator (CEA) 62, 73, 84, 87, 88, 123, 124, 176
cargo inspection 10
Carnegie Institute, Washington 6
Cavendish Laboratory 1, 3
cavities, superconducting 95
cavity, accelerating 31, 58, 90, 94, 144
Cerenkov counter 110
Cerenkov detector 112
Cerenkov, Pavel Alekseyevich 112
CERN 63, 64
CESAR (CERN Electron Storage and Accumulation Ring) 95
Charpak, Georges 115
Chodorow, Marvin 35
Christofilos, Nicholas C. 44, 61, 63, 171
Clayton, C. 163
CLIC (Compact Linear Collider) 143, 146, 148, 173
    Test Facility 3 147
cloud chamber 112, 113
Cockcroft, Sir John Douglas 2
Cockcroft-Walton 1–5
    accelerator 4
collider 78–109
    asymmetric ring 97
    electron–electron 81
    electron–positron 81

collider
    linear 81, 104
    proton-antiproton 94
    proton–proton 91
    single-ring 84
Collins, Thomas 73
collisions, head-on 78, 79
Condon, E. 2
cooling 96, 154–157
    electron 155
    ionisation 154
    stochastic 99–101
cosmotron 58–63
Coulomb barrier 4
Courant, Ernest D. 62
CP violation 103
Crystal Ball Detector 111
CSNS 150
cyclotron 82
    11-inch 14
    184-inch 20
    60-inch cyclotron 15
    applications 26
    cancer therapy 134
    early 14
    modern 26
    permanent magnets 27
    superconducting 153

D

Dahl, Odd 6
damping 92
    ring, ILC 145
DARHT (Dual Axis Radiological Hydrodynamic Test Facility) 47
Dawson, John Myrick 163
de Broglie, Louis-Victor xiii, 5
Dees, cyclotron 19
Delahaye, J.-P. 147
DESY (Deutsches Elektronen-Synchrotron) 88
    aerial 102
detector, modern 115
    early 110
    for synchrotron radiation sources 119
diffraction of synchrotron light 115, 120
diffraction, neutron 149
digital x-ray imaging 117
DORIS 84, 88, 97
Doubler 70, 72, 73, 75
drift tubes 12
Dubna 42, 45

E

École Polytechnique 161, 163, 165
Edwards, Helen 77

Elder, F.R. 171
electron-electron storage ring
    Stanford 83
    VEP1 85
electron-positron storage ring
    AdA 85
    ADONE 87
electron microscope 10
electron synchrotron, early 57
electrostatic accelerator 1–9, 169
    applications of 9
    commercial production 9
emittance 39, 103, 137
emulsion micrograph 112
emulsion, particle tracks in 116
energy 158
    amplifier 99
    doubler (see Doubler)
ERA (Electron Ring Accelerator) 3, 30, 67, 107, 114
ESRF (European Synchrotron Radiation Facility) 127
ETOILE 138
European Union X-Ray Free Electron Laser 128
Evans, Lyn 107
extraction 51, 69, 73, 77, 83

F

FAIR (Facility for Antiproton and Ion Research) 151, 152
Fang, Shouxian 167
Farley, F 93
Fermi National Accelerator Laboratory 70
    aerial view 71
    high-rise 71
FFAG 22, 91, 97, 136, 156
FLASH 128
focusing 18, 19, 22, 30, 31, 34, 60–63
    strong 61
    transverse 8
    weak 58
Frascati (INFN) 81, 84, 85, 93
FEL (Free Electron Laser) 41, 129, 131
FXR induction accelerator 46

G

g-2 experiment 100
Gabor, Denis 11
Gaede, Wolfgang 12
Gamma Ray Knife 134
Gamow, George 2, 3, 4
gantry 135, 136, 138
Geiger, Hans 3
Geiger-Muller counter 110
Ghiorso, Albert 42
Ginzton, Edward L. 35–38
Glaser, Donald 113
glioblastoma 138

Goward, F.K. 54, 56, 57
Green, George Kenneth 67
Grunder, Hermann A. 41, 42
GSI (Gesellshaft für Schwerionen) 137, 138, 151, 152
Gurney, R.W. 2

H
hadrons vs. leptons 78
hadron therapy 136, 138
Halbach, Klaus 124, 125
Hansen, William W. 37
Harvard cyclotron 134, 135, 137
Harvard Medical School 10
HASYLAB 88
Havens, G.G. 6
heavy-ion fusion xv, 157, 159
Heavy Ion Virtual National Laboratory 158
heavy-ion colliders 107
HEDP (high energy density plasmas) 157, 159
Helios 76
HERA (Hadron Electron Ring Accelerator) 97, 102
Hereward, Hugh 64, 69
Herb, Raymond 8
Hertz, H. 122
HICAT (Heidelberg Ion Therapy Centre) 138
Higgs particle 106
High Voltage Engineering Co. 9
HIMAC (Heavy Ion Medical Acceleratory Facility) 137
Hine, Mervyn 64, 68, 69
Hofmann, Albert J. 79
Hübner, Kurt 149
Hulse, Russell A. 123

I
IBA 135, 138
IBX (Integrated Beam Experiment) 158
ICECUBE 117
ICFA (International Committee for Future Accelerators) 141, 149
IFMIF (International Fusion Materials Irradiation Facility) 159
ILC (International Linear Collider) 144
    tests 164
induction linacs, applications 44–46
inertial confinement 157
inverse free electron laser 162
ion accelerators xv, 16, 18, 43, 166
ion implantation 9, 34, 140
ion therapy 55, 138
ISABELLE 98
Ising, Gustav 1, 28–31, 34, 44
ISIS 150
ISOLDE 151
isotope production 9, 26, 34
ISR (Intersecting Storage Ring) 91–99

ITER (International Thermonuclear Experimental Reactor) 158, 159

J
JAERI, Tokay, Japan 150
Jassinsky, W.W. 51
Jefferson Laboratory 41, 42
JLC (Japanese Linear Collider) 143, 146
Johnsen, Kjell 97
Joshi, Chandrashekhar 164

K
Kamiokande 152
KAMLAND 117
Kapchinskii, I.M. 34
Kapitza, Peter 2, 3
Karolinska Institute 138
Katsouleas, Thomas 163
Keil, E. 93
Kerst, Donald W. 52
keV 2–4, 29, 30, 34
klystron 36
Kolomenski, A.A. 23
Krienen, F. 93
Kulipanov, G.N. 156
Kurokawa, Shin-ichi 104

L
LAMPF (Los Alamos Meson Physics Facility) 33
Landau damping 34, 69, 178
laser accelerator 162, 172
    self-modulated 165
Laslett, Jackson 15
lattice 67, 124
Lawrence Berkeley National Laboratory 16, 17
Lawrence, Ernest O. 12, 170
Lawson, John 162
LCLS (Linac Coherent Light Source) 38, 40, 131
LEP 78, 86
LHC (Large Hadron Collider) 103
    aerial view 90
lasers and plasmas 161
Lichtenberg, D.B. 81
light ions, for therapy 137, 138
light source, fourth generation 132
linac 28–46
    Alvarez 32
    cancer therapy 135
    electron 34
    for driving a reactor 160
    for therapy 134, 135, 136, 138
    heavy ion 42
    induction 44
    proton 30
linear accelerator, early 30

linear collider 141, 143–146
Liouville theorem 155
Livingston, Milton S. 13, 170
Lofgren, Edward 60
Loew, Gregory A. 38, 39
Loma Linda 136, 137
luminosity 80
LUX 132, 133
LWFA (Laser Wakefield Accelerator) 163

M

Madey, John 129
magnets, superconducting 75
matrix 76, 105, 119
Maxwell, J.C. 122
McDaniel, Boyce D. 66
McMillan, E.M. 18
Med-Austron 138
MICE (Muon Ionization Cooling Experiment) 154, 157
Midwest Proton Radiotherapy Center 137
Miller, Richard 26
Miller, Roger Heering 40
MIT 6–9
MTA (Materials Testing Accelerator) 32, 33
multiwire proportional chambers 115
muon colliders 76, 151, 153–156
muon storage rings (g-2) 93
MURA (Midwestern Universities Research Association) 23

N

National Electrostatic Corporation 9
neutrino factory 153, 154, 156, 157
neutrino super beams 153
neutron diffraction 150
neutron spallation source 48
neutron therapy 135
Newton, R.G. 81
NIF (National Ignition Facility) 157, 159
Nimrod 360
niobium 90, 92, 94
NLC (Next Linear Collider) 143
Northeast Proton Therapy Center 137
Novalis 134
nuclear reaction, first man-made 4, 8
nuclear waste disposal 30
Nygren, David 117

O

Oddone, Pier 103
Ohkawa, Tihiro 25
Oide, Katsunobu 104
Oliphant, Marcus Laurence Elwin 56
O'Neill, Gerard K. 82
Omnitron 42
Optivus 135

Orsay 81, 84, 86
Ozaki, Satoshi 108

P

Palmer, Robert B. 154
Panofsky, W.K.H. 38
pattern recognition programs 113
Pauly, T. 9
pbar 105
PBWA (Plasma Beat Wave Accelerator) 163
Pellegrini, Claudio 130
Pelletrons 8, 9
Penny, G.W. 51
PEP 97
PEP-II 38, 40
PETRA 84
Petrucci, G. 93
phase focusing 18, 22
phase stability 60, 63
Phasotron 60
photomultipliers 110
Picasso, Emilio 93
PIMMS (Proton and Ion Medical Machine Study) 138
PIXIE 10
ponderomotive potential (sp) 162
protein molecule 122
proton drivers for power reactors 160
proton synchrotron 56, 57
    1 GeV, Birmingham University 58
    early 56
proton therapy 66, 135, 136, 138
psi 40, 82, 124, 135, 138, 173
PWFA (Plasma Wakefield Accelerator) 163

Q

quadrupole 33, 34, 69
quark-gluon plasma 107–109

R

Rad 16, 18, 21
radioactive waste 160
radioisotope, production of 125
Radionics and Electra Inc. 134
RAL (Rutherford Appleton Laboratory) 165, 173
Rare Isotope Accelerator 151
Rayleigh length 165
ray transformer. See betatron
reactors, accelerator driven 145, 150
Reardon, Paul 72
recirculated beam 41
Rees, John Robert 87
relativistic limit 19, 23
RFQ (Radio Frequency Quadrupole) 34, 35
RHIC (Relativistic Heavy Ion Collider) 108, 109
RIA (Rare Ion Accelerator) 152

RIBF (Radioactive Isotope Beam Facility)   151
Richter, Burton   82
RIKEN   42
Robinson, Kenneth   80
Roentgen, W.C.   49
Ross, H.M.   81
Round Hill   7
Rubbia, Carlo   99
Rutherford, Ernest   xii

S

Sands, Mathew   91
SASE (Self Amplified Spontaneous Emission)   128
Scherrer, Paul   26
Schiff, Leonard   19, 21
Schnell, Wolfgang   148
Schopper, Herwig   88, 92
Schott, G.A
Schwettman, H.A.   41
Schwinger, Julian   123
scintillators   110
SCSS (SPring-8 Compact SASE Source)   128, 130
Seaborg, Glenn   15, 18
Seeman, John   103
semiconductor production   10
Serber, Robert   19, 51, 171
SESAME   81, 126
Siemens Co   11
silk belts   5
Skrinsky, Alexander   156
SLAC (Stanford Linear Accelerator Center)   38, 39
SLC (Stanford Linear Collider)   86
Slepian, J.   50
Sloan, David   134
Smith, T.   41
SMLWFA (Self-Modulated Laser Wakefield Accelerator)   165
Smokatron   45, 177
smooth approximation   25
SNS (Spallation Neutron Source)   151
Snyder, Hartland S.   62
Solvay Conference   17
space charge   34
spallation neutron sources   148, 151
spark chamber   115
SPEAR   38, 39, 40
SPEAR 3   128
spiral sector cyclotrons   23
SPring-8   127
SPS (Super Proton Synchrotron)   73
stacking   23, 91, 95, 97, 141, 148, 182
Standard Model   105
Stanford Mark III   37
Stanford-Princeton Collider   81
Steenbeck, Max   11, 51
stochastic cooling   100

pickup   101
principle   101
storage rings (*see* colliders)
strong focusing   60
superconducting accelerating cavities   94
superconducting magnets   74
SuperHILAC   42
swindletron   9
Symon, Keith R.   22, 23, 25, 91
synchro-phasotron   60, 61
synchrotron   54–64
    Birmingham   58
    CERN PS   33, 64, 65, 91, 148
    electron   56
    first   56
    FNAL   67, 103, 104, 172, 186, 187
    for cancer therapy   136
    proof of principal betatron   57
synchrotron radiation   122
    applications   126
    emission   123
    first observation of   123
    radiation research   43, 173
Superconducting Super Collider (SSC)   98
synchrotron radiation sources
    first generation   123
    fourth generation   132
    second generation   123
    third generation   124
Szilard, Leo   11, 51

T

Telnov, Valery Ivanovich   146, 147
Teplyakov, V.A.   34
TESLA   146
Tevatron   97, 153
TFT (thin film transistor)   119
TFTR   158
Thomas, Llewellyn H.   20
thorium   160
thyratron   44
Tigner, Maury   89
Tollestrup, Alvin W.   76
Touschek, B.   81, 86, 172
TPC (Time Projection Chamber)   116–118
transition   xi, 64, 97, 148
transuranic   42, 160
treatment table   136
TRISTAN   104, 108
TRIUMF   26
Tuck, James L.   51
Tuve, Merle   6, 50, 51

U

undulator   124, 125, 128, 129, 131

UNESCO 64
University of Wisconsin 6, 8, 9
uranium, separation 12, 21

V

van der Meer, Simon 99
Van de Graaff accelerator 6
Van de Graaff, Robert Jemison 5–10, 134
Van de Graaff, tandem 5, 8
Varian Medical Systems Inc. 136
Vasa 122
VEP1 81, 84, 85
vertex detector 116, 119
Veksler, Vladimir Iosifovich 22
Vinokurov, N.A. 156
voltage multiplying columns 4
Voss, Gustav-Adolf 81

W

Walton, Ernest T.S. 2
Westinghouse Electric Company 50

Wideroe, Rolf 30
wigglers 124, 125
Wiik, Bjorn 86, 88, 89
Wilson, C.T.R. 110, 112
Wilson, Ian 147
Wilson, Martin 76
Wilson, Robert Rathbun 66
Winick, Herman 125
wire chamber 115
Woodyard-Lawson theorem 161
Woolwich Arsenal Research Laboratory 56, 57
W and *Z* discovery 64, 98, 105, 141, 155

X

X-Band Test Accelerator 144
x-ray 49–51, 53, 54

Z

ZGS (Zero-Gradient Synchrotron) 22, 23
Zilverschoon, K. 93, 97